Theoretical and Applied Mathematics

Earth Sciences in the 21st Century

nova
science publishers
New York

Theoretical and Applied Mathematics

Applied Computational Hydraulics for Engineers
Necati Ağıralioğlu and Mehmet Özger
2024. ISBN: 979-8-89113-415-7 (Hardcover)
2024. ISBN: 979-8-89113-506-2 (eBook)

The Impossible Quintic Made as Simple as Possible
David Abraham Kault, Graeme Sneddon and Sam Kault
2023. ISBN: 979-8-88697-908-4 (Softcover)
2023. ISBN: 979-8-88697-972-5 (eBook)

More information about this series can be found at
https://novapublishers.com/product-category/series/theoretical-and-applied-mathematics/

Earth Sciences in the 21st Century

Neogene Deep Water Benthic Foraminifera from the Indian Ocean – A Monograph
Anil Kumar Gupta
2022. ISBN: 979-8-88697-385-3 (Hardcover)
2022. ISBN: 979-8-88697-412-6 (eBook)

Observing Micrometeorology: A Personal Tour through an Evolving Science
Bruce B. Hicks
2022. ISBN: 979-8-88697-369-3 (Hardcover)
2022. ISBN: 979-8-88697-426-3 (eBook)

More information about this series can be found at
https://novapublishers.com/product-category/series/earth-sciences-in-the-21st-century/

Necati Ağıralioğlu
and Mehmet Özger

Applied Computational Hydraulics for Engineers

nova
science publishers
New York

DOI: https://doi.org/10.52305/PWFR7002

NOTICE TO THE READER

The Publisher has taken reasonable care in the preparation of this book, but makes no expressed or implied warranty of any kind and assumes no responsibility for any errors or omissions. No liability is assumed for incidental or consequential damages in connection with or arising out of information contained in this book. The Publisher shall not be liable for any special, consequential, or exemplary damages resulting, in whole or in part, from the readers' use of, or reliance upon, this material. Any parts of this book based on government reports are so indicated and copyright is claimed for those parts to the extent applicable to compilations of such works.

Independent verification should be sought for any data, advice or recommendations contained in this book. In addition, no responsibility is assumed by the Publisher for any injury and/or damage to persons or property arising from any methods, products, instructions, ideas or otherwise contained in this publication.

This publication is designed to provide accurate and authoritative information with regard to the subject matter covered herein. It is sold with the clear understanding that the Publisher is not engaged in rendering legal or any other professional services. If legal or any other expert assistance is required, the services of a competent person should be sought. FROM A DECLARATION OF PARTICIPANTS JOINTLY ADOPTED BY A COMMITTEE OF THE AMERICAN BAR ASSOCIATION AND A COMMITTEE OF PUBLISHERS.

Additional color graphics may be available in the e-book version of this book.

Library of Congress Cataloging-in-Publication Data

ISBN: 979-8-89113-415-7

Published by Nova Science Publishers, Inc. † New York

Contents

Chapter 1

Introduction to Computational Hydraulics

1.1. Definition of Computational Hydraulics

Computational Hydraulics is an applied science that tries to solve many water flow related processes by simulation. In this branch, there is a method that solves the problems with the usage of computers (Szymkiewicz et al., 2016). Computational method is between pure theoretical and experimental methods and is closer to the experimental method. Computational Hydraulics simulates water flow using numerical methods in computers. Actually, there is no significant difference between computational hydraulics and computational hydrodynamic or computational fluid dynamics. But in the other two terms, the issues are more limited to fluid. Some typical practical problems in the hydraulics are rarely related to direct flow. Computational Hydraulics is more concerned with its consequences than the flow itself. Similar techniques are used to overcome issues with forces coming to a sheet pile, seepage through an earthen dam body, and heat transmission. The title "Applied Computational Hydraulics" is picked for the book since it solves numerous practical issues that are comparable to one another.

The broad concepts of computational hydraulics, the structure of the general equations of the water flow equations, the primary discretization techniques, and definitions of modeling and hydraulic modeling are all covered in this introduction.

1.2. Basic Principles of Computational Hydraulics

The principles of computational hydraulics described in this book also can be easily used for solving civil engineering as well as coastal engineering and environmental engineering problems.

The approximation of a continuously changing variable according to its values over a finite number of points is called discretization. As an example, the corresponding values to a continuous parabola curve, in Figure 1.1.a, can be shown by its discretized values in Figure 1.1.b.

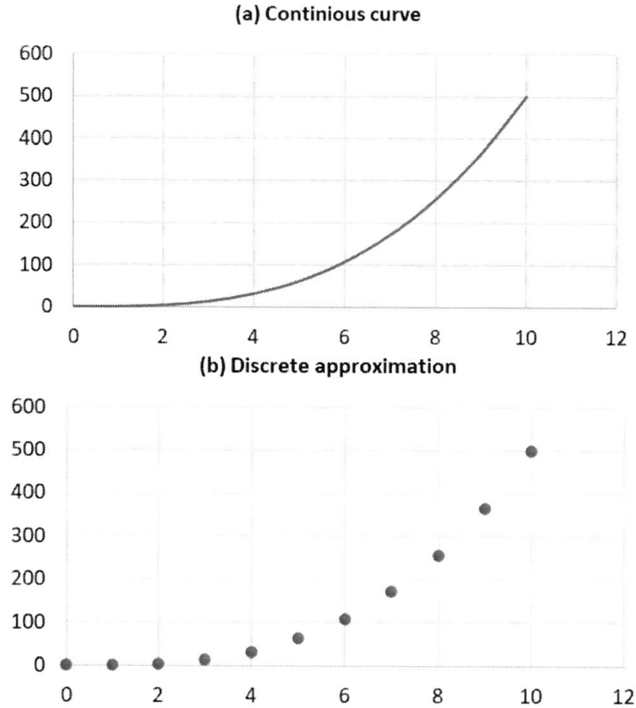

Figure 1.1. Continuous curve and its discrete approximation.

The main elements of any computational hydraulics simulation are:

1. The flow field is discretized. Thus, the field variables such as velocity, acceleration and pressure are approximated by the values in terms of finite number of points.
2. The equations of motion are discretized in terms of values at points. Thus, differential equations turn into algebraic equations which are easy to solve.
3. The resulting set of algebraic equations is solved to find their values at the corresponding discretized points.

The main steps in any computational hydraulics simulation are:

1. Pre- processing:
   a. Formulation of the problem involving the determination of basic equations and boundary conditions,

   b. Setting up the calculation mesh.
2. Solution:
   a. Discretization of basic equations,
   b. Solution of the resulting algebraic equations.
3. Post-processing:
   a. Displaying the graphics,
   b. Evaluation of the results.

1.3. Forms of the Governing Water-Flow Equations

The equations of fluid flow are based on basic physical principles and they are:

1. Conservation of mass (continuity equation): The mass of fluid within a control volume does not change. This means that the mass will not change physically within the control volume as can be seen in Figure 1.2. The conservation of mass equation is sometimes called as continuity equation in hydraulics.
2. Conservation of momentum: Force multiplied by time equals momentum. The concept of conservation of momentum states that the overall momentum of a closed system cannot be altered while a control volume is not being affected by any external force. Navier-Stokes equations are conservation of momentum equations for compressible, viscous flow states.

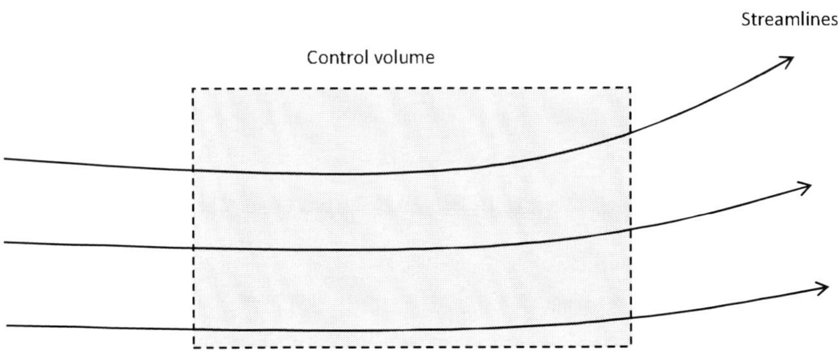

Figure 1.2. Control volume for a fluid flow.

3. Conservation of energy: Energy can be converted from one form to another, however, in a given control volume, the total energy remains constant. This is called the principle of conservation of energy.

Basic governing equations used in hydraulics will be summarized in Chapter 2.

1.4. Integral Equations

In integral equations, an unknown function is appeared under an integral sign. There is a close link between integral equations and differential equations. Some problems can be solved in either way. In the solution of the integral equation, a control volume is taken into account. In this approach, changes in the total amount of physical quantities such as mass, momentum or energy within the control volume are considered.

The balance of a physical quantity over a given time interval for any control volume is written as follows:

Amount of change = Incoming amount into volume – Outgoing amount from volume + Amount produced (1.1)

For instance, the conservation of mass could be expressed as:

Mass entering per unit time = mass leaving per unit time + mass increase in the control volume per unit time for any control volume.

The balance equation in hydraulics is expressed in terms of amount passing through the control volume per unit time. Thus, the terms in the equation of balance are divided by the time and the equation of balance is written as follows:

Rate of change (inside volume) + Net flux (out of boundary) = Source (inside volume) (1.2)

Within the control volume, there is no increase in rate (or discharge) for steady flow. So, it is possible to write:

Rate entering = Rate leaving (1.3)

The transport rate or flux passing through a surface consists of advection and diffusion. The former moves with the flow of fluid, the latter is the net transport with random molecules or turbulence. So, the balance equation can be written as follows:

Rate of change (inside volume) + Advection + Diffusion (out of boundary) = Source (inside volume) (1.4)

It is very important that the equation is single generic equation regardless of the physical quantity of interest, such as mass, momentum, and chemical compound. This indicates that the equation has only one variable, such as water volume, and can be used for general contents such as water volume, pollutant concentration.

We can consider an example of conservation equation in integral form. For this purpose, a control volume for an open channel flow which is a hypothetic tool for defining the various fluxes and forces can be used. A conceptual control volume for open-channel flow is shown in Figure 1.3.

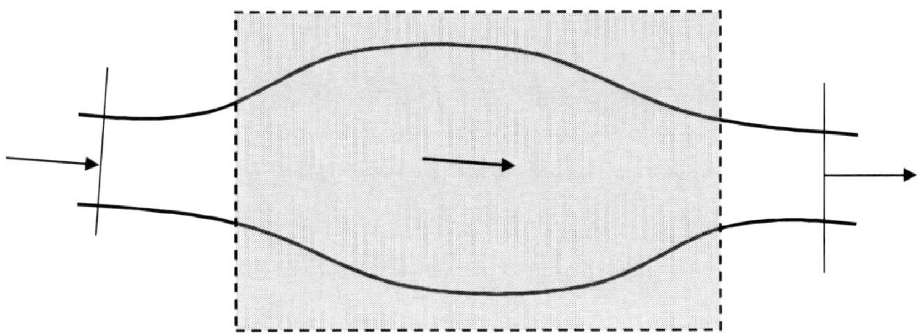

Figure 1.3. Control volume for a stream (plan view).

The conservation of mass principle for a control volume can be written as

$$\int_{x_L}^{x_R} [A(x, t_U) - A(x, t_D)]dx = \int_{t_D}^{x_u} [Q(x_L, t) + I(t) - Q(x_R, t)]dt \quad (1.5)$$

where, A is the wetted cross-sectional area,

Q is the flowrate,

I is the lateral inflow,

x_L is the upstream boundary of the control volume,

x_R is the downstream boundary of the control volume,
t_U is the initial time,
t_D is the, one-time step after t_U,
x is the distance,
t is the time.

This is a 1-D equation for conservation of mass in integral form. We can solve it using any method as well as numerical methods. In the control volume, we can explain Equation (1.5) as

Change of volume = Volume of inflow - Volume of outflow + Lateral inflow (1.6)

1.5. Differential Equations

Some ordinary or partial differential equations are used to represent many problems in engineering as well as in hydraulics. Fluid flow equations can also be written in an equivalent manner in differential form. In this form, the development of the problem at one point is described, not over the entire control volume.

As an example of a partial differential equation, we can consider the advection-diffusion equation applicable to variable cross sections.

$$\frac{\partial(AC)}{\partial t} = \frac{\partial}{\partial x}\left(D\frac{\partial C}{\partial x}\right) - \frac{\partial(VAC)}{\partial x} \tag{1.7}$$

C is the concentration, A is area, V is velocity and D is diffusion coefficient. Equation (1.7) is also another kind of mass conservation equation.

If A is non-variable, we can write Equation (1.7) as

$$\frac{\partial C}{\partial t} + a\frac{\partial C}{\partial x} = \nu\frac{\partial^2 C}{\partial x^2} \tag{1.8}$$

where, a is constant as advection velocity, ν is the kinematic viscosity.

The term ($\partial C/\partial x$) is called the advective or convective term, the ($\partial^2 C/\partial x^2$) is called diffusive or viscous term. Equation (1.8) is also referred to as one dimensional advection- diffusion equation or one dimensional convection –diffusion equation.

As an example of partial differential equation, we can write one dimensional conservation of mass equation, as a partial differential equation, for water flow as,

$$\frac{\partial A}{\partial t} + \frac{\partial Q}{\partial x} = q \tag{1.9}$$

where, A is the cross sectional area,

Q is the discharge,

q is the lateral inflow,

x is the space,

t is the time.

This equation is also called continuity equation.

As an example of the ordinary differential equation, the basic equation is given for a reservoir of the dam as

$$\frac{dS}{dt} = I - O \tag{1.10}$$

in which;

I is the inflow to reservoir,

0 is the outflow from reservoir,

S is the reservoir storage,

dS/dt is the rate of change of storage within a reservoir. This equation is called as continuity equation (see Figure 1.4).

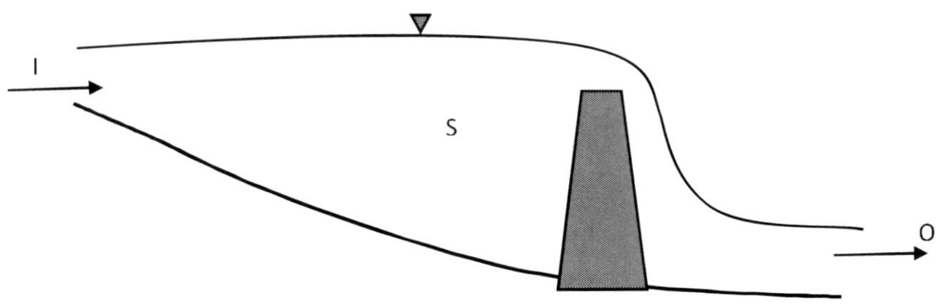

Figure 1.4. Definition of continuity equation for a reservoir.

If a steady flow exists, namely dS/dt = 0, Equation (1.10) becomes I = O = Constant. In this case, considering Figure 1.5, we can write the continuity equation as:

$$Q = A_1V_1 = A_2V_2 \tag{1.11}$$

where Q is the volumetric flow rate,
A is the cross-sectional area of flow,
V is the mean velocity.

Figure 1.5. Continuity equation for steady water flow.

1.6. Numerical Methods of Ordinary Differential Equations

Numerical methods, which are sometimes called numerical integration, are also used to find numerical approximations to the solutions of ordinary or partial differential equations.

We can classify the numerical methods used in solution of ordinary differential equations;

- Single-Step Methods for Initial Value Problems: a. Euler, b. Heun and Midpoint/Improved Polygon, c. General Runge-Kutta, d. Adaptive step-size control,
- Stiff Ordinary Differential Equations,
- Multi-Step Methods for Initial Value Problems: a. Non-Self-Starting Heun, b. Newton-Cotes, c. Adams,
- Boundary Value Problems: a. Solution Methods, b. Shooting Method c. Finite Difference Method, d. Eigenvalue Problems.

In this section, we want to consider the Euler Method and the Modified Euler Method in detail used commonly in some practical solutions of the ordinary differential equations (Gentry et al., 1966).

1.6.1. The Euler Method

We start to consider a general first order initial value problem, for initial condition $y(t_o) = y_0$,

$$\frac{dy}{dt} = f(t, y) \quad (1.12)$$

where f(t, y) is a known function and the values in the initial condition are also known numbers. We want to solve Equation (1.13) near $t = t_0$.

Euler method is one of the simplest numerical methods. In order to derive Euler method, we consider the standard derivative approximation for any function y(t) as,

$$y'(t) \approx \frac{y_{n+1} - y_n}{t_{n+1} - t_n} = \frac{y_{n+1} - y_n}{h} = f(t, y) \quad (1.13)$$

where h is the step size. This is a forward difference approximation in finite difference method. In fact, Explicit Euler scheme is based on a first order Taylor approximation to the solution. Taylor approximation is further discussed in Chapter 5.

For approximating solutions of first-order initial-value problems we can write, $y' = f(t, y)$, $y(t_0) = y_0$.

Considering Figure 1.6, the slope at (t_0, y_0) is $y' = f(t_0, y_0)$.

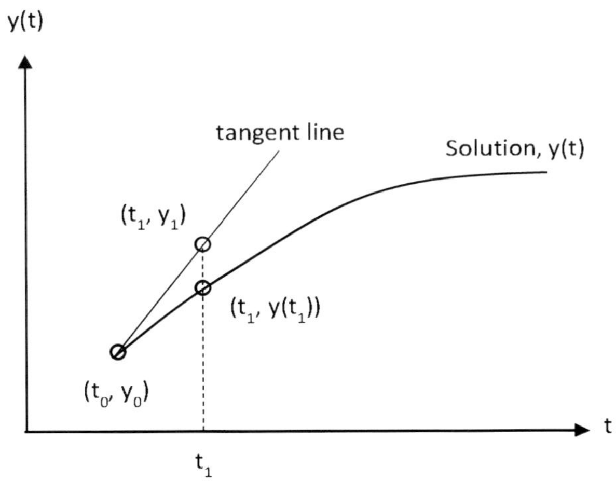

Figure 1.6. An illustration for Euler method.

As it can be seen from Figure 1.6, the approximate value of the solution when $t = t_1$ is

$$y_1 = y_o + \frac{f(t_0, y_0)}{t_1 - t_0} \tag{1.14}$$

for the solution at equally spaced numbers similarly we can write,

$$y_2 = y_1 + \frac{f(t_1, y_1)}{t_2 - t_1} \tag{1.15}$$

In general,

$$y_{n+1} = y_n + \frac{f(t_n, y_n)}{t_{n+1} - t_n} \tag{1.16}$$

Taking $t_{n+1} - t_n = h$, and $f(t_n, y_n) = f_n$, we can write

$$y_{n+1} = y_n + hf_n \tag{1.17}$$

This equation can be obtained from Equation (1.12). From the boundary condition y_0 is almost known. Equation (1.15) gives a rule to compute y_1, y_2, ..., y_{n+1} successively. This approach is employed by the most of numerical methods for solving ordinary differential equations. Euler method is an explicit method. In this method, the global error is O(h).

1.6.2. The Modified Euler Method

The modified Euler method begins with an Euler step, which provides a temporary value for y_{n+1} at the following step t_{n+1}.

$$y_{n+1} = y_n + hf(t_n, y_n) \tag{1.18}$$

This step is similar to the Euler step, but instead of y, it uses the average of y at the temporary point and y at the step's beginning point:

$$y_{n+1} = y_n + \frac{h}{2}[f(t_n, y_n) + f(t_{n+1}, y_{n+1})] \tag{1.19}$$

y_{n+1} is unknown in this case. However, we may estimate it using the Euler technique, which yields a two-stage predictor-corrector algorithm known as the Modified Euler-algorithm, as follows:

Step 1: Predictor;

$$y'_{n+1} = y_n + hf(t_n + y_n) \tag{1.20}$$

Step 2: Corrector;

$$y_{n+1} = y_n + \frac{h}{2}[f(t_n, y_n) + f(t_{n+1}, y'_{n+1})] \tag{1.21}$$

In this method, the order of accuracy is $O(h^3)$ per step.

1.7. Numerical Methods of Partial Differential Equations

Numerical methods used in the solution of partial differential equations can be classified as (Szymkiewicz, Huang and Szymkiewicz, 2016):

- Finite Difference Method,
- Characteristics Method,
- Finite Element Method,
- Finite Volume Method,
- Boundary Element Method.

Finite Difference Method: The finite difference approach converts the partial differential equations that express the problem into finite difference equivalents, which are algebraic equations. The beginning and/or boundary conditions required for problem solving are then determined. Then, using computers, finite difference expressions, which are algebraic statements, are solved. The partial differential equation used to solve the problem can be elliptical parabolic, or hyperbolic. Chapter 5 explains the general principles of the finite difference approach. The beginning and boundary conditions of the problems are given in Chapter 6. In Chapters 7, 8, and 9, solutions to elliptic, parabolic, and hyperbolic partial differential equations using finite difference methods are presented, along with engineering applications.

Characteristics Method: The characteristics method is a technique used in the solution of partial differential equations. The method of characteristics is more generally applicable to the solution of any hyperbolic differential equation. Using this method, partial differential equations become ordinary differential equations. Then these ordinary differential equations can be solved either by analytical or numerical methods. The characteristics method and some of its engineering applications are explained in Chapter 12.

Finite Element Method: In finite element method, the solution area is divided into elements. Then these elements are connected with some nodes. These elements can be triangular in two-dimensional problems and tetrahedral in three-dimensional problems. Depending on these elements, the boundary conditions are determined at the boundary points, so that the total number of points is obtained according to the points on the border and the inner points. At the end of this process, continuous medium equations turn into some simultaneous algebraic equations. The total number of points multiplied by the number of unknowns gives the degree of freedom. Finite element method can be used in all kinds of boundaries in the solution of general and special equations in engineering and other physics applications. This method is commonly used especially in the solution of elliptic and parabolic type partial differential equations. The disadvantage of this method is its complexity.

Finite Volume Method: The finite volume method is a discretization method used for the approximate solution of a single or a system of partial differential equations. In this method, it creates a mesh that divides the solution domain into parts. These elements formed by this mesh are called control volumes. The integration of the partial differential equation on each control volume gives a balance equation. This set of balance equations is discretized. The resulting discrete equations are based on the discrete number of unknowns and can be linear or nonlinear according to the partial differential equation.

Boundary Element Method: In a boundary value problem analysis, solving boundary integral equations is a classic way to solve partial differential equations. As it is understood from its name, the boundary element method means solving boundary integral equations with a numerical approach method. This method only needs discretization at its boundaries. There is also a solution only at the boundaries. However, mathematical operations are more difficult than finite difference or finite element methods in solving boundary integral equations. Also, if there are irregular shapes in the boundaries, it may create a singularity in the solutions.

1.8. Terms of Hydraulic Modeling

Researchers proposed a list of terms for study in an effort to create a common language among individuals working on models for hydraulic phenomena. The glossary is divided into five sections: Model types in general, scaling and design of hydraulic models, mathematical model types, and words pertaining to numerical models are all covered by ASCE (1982).

Some of the general terms in modeling are given below (ASCE, 1982):

- Adjustment: The process of fitting a model to ensure that it accurately reproduces a set of prototype conditions.
- Calibration: the process of fitting or standardizing the operating characteristics of instruments and model accessories to a physical model or coefficients in a mathematical model. The process of evaluating the scale values of an instrument with respect to the physical quantity being measured.
- Hydraulic model: Physical model that uses water as a fluid.
- Mathematical model: a mathematical model is a description of a system using mathematical relationships and concepts.
- Model: a reproduction of the prototype, usually on a small scale, but it may be larger or geometrically distorted.
- Physical model: a model is a smaller physical copy of an object using the physical properties of modeling materials.
- Prototype: an original full-scale model of the structure, system process, or phenomenon being modeled.
- Simulation: is a model or representative example of the prototype.
- Steady state model: a model in which the variables under study does not change with time.
- Unsteady state model: a model in which the variables under study are change with time.
- Validation: the process of checking the accuracy of the techniques by comparing the model results with a set of prototype data not used for verification.
- Verification: confirming the accuracy of the fitted model against the prototype model.

1.8.1. Terms Referring to Numerical Modeling

The terms of numerical modeling are explained below (ASCE, 1982):

- Accuracy: the closeness of the results computed by a numerical solution to the exact solution of the fundamental equations.
- Boundary conditions: Boundary conditions are constraints necessary to solve a problem.
- Characteristic method: in the characteristic method, the governing partial differential equations are transformed into characteristic equations in a mathematical model and these equations are then solved using characteristic curves.
- Consistency: Quality of a discrete representation that behaves like the continuum equations and solution as the discretization becomes finer.
- Convergence: Convergence is the property of tending to a unique solution in mathematical modeling.
- Discretization: Discretization is the process of converting the representation of a continuous variable into discrete values at specific points in space and/or time.
- Discretization error: discretization error is the error introduced by the discrete representation of a continuous variable.
- Explicit scheme: the explicit scheme is the scheme in which independent values in the numerical modeling of fundamental equations are found directly and explicitly from the previously known values.
- Finite difference method: The method of solving the governing equations of a mathematical model is approximated by finite intervals in the dependent variable.
- Finite element method: Finite element method is a numerical method in which the spatial domain is divided into elements and the governing equations on each element are solved approximately.
- Grid: Grid of points covering the space or time-space domain of a numerical model.
- Implicit scheme: implicit scheme is the scheme where independent values in numerical modeling of fundamental equations are not found directly and explicitly from the previously known values.

- Initial conditions: An arbitrary set of initial values that belong to or are imposed on the given values of the dependent variables in an equation
- Numerical diffusivity: Numerical diffusivity refers to the artificial diffusion introduced in numerical methods when approximating solutions to partial differential equations.
- Numerical Dispersion: Numerical smearing or smoothing of a solution resulting from numerical diffusivity.
- Rounding Error: Rounding error is the cumulative error that results from rounding results, since only a finite number of digits remain after each arithmetic operation.
- Stability: Stability refers to numerical schemes that produce bounded solution errors with respect to differential equations based on the approximation scheme used.
- Truncation error: This error is the error arising from the use of truncated Taylor series expansions by substituting the finite difference equivalents of differential equations.

Chapter 2

Review of Governing Equations

2.1. Introduction

Some basic equations and their simplified versions are used to study many topics in hydraulics. Fundamental equations can be written as integral or differential equations. Since numerical methods are used to seek solutions in this book, the differential forms of the general equations are explained.

These differential equations include one-, two-, and three-dimensional partial differential equations based on time and space. Partial differential equations in hydraulics can be used in the representation of problems involving pressurized pipes, open channels, underground flows, and the transport of various materials through air or water.

Fundamental equations in the form of partial differential equations are used not only in hydraulics, but also in fluid mechanics and many engineering disciplines. They are also useful in explaining many processes in mechanical engineering, aircraft engineering, marine engineering, chemical engineering, and petroleum engineering. They are even the fundamental equations in many applied branches of science such as biology. For example, Laplace's equation applies to all potential flows in hydraulics, to some heat transfer problems, to structural analysis, to electric fields, and to many other fields. The formula considered is the same, but the solution in each case depends on the boundary conditions that define the physics of the event. Although the same equation is used for all problems, it is used to understand many different problems depending on the boundary conditions.

The partial differential equations presented here can be used to solve many problems, such as. Wave propagation, flows in pipes or open channels, transport and distribution of matter in water and air, heat transport, vibrations in solid mechanics, stress-strain relationships in structures and machine parts, electric fields and potentials, diffusion of chemicals in water and air, electromagnetic fields and quantum mechanics, heating and cooling problems, advection or diffusion of moisture, evaporation, air movements, frictional forces, evaporation.

In this section, the equations used in hydraulics and some other engineering fields are not derived, but only the differential form of the equations is explained. The three basic equations used in solving the problems in hydraulics as well as in other applied physical sciences are based on the principles of physics such as,

1. The principle of conservation of mass
2. The principle of conservation of momentum and
3. The principle of conservation of energy.

The continuity equation, the Navier-Stokes equation, and the energy equation, which were created according to these three principles, are explained here in both Cartesian and cylindrical coordinates.

2.2. Continuity Equation

2.2.1. General Remarks

The principle of conservation of mass states that the mass of a body does not change during its motion. This can also be expressed in terms of the mass of a body changing over time. The principle of conservation of mass should be expressed mathematically for any substance. This rule is also valid in physics, chemistry and biology.

For example, this law applies to the condensation of steam. In such a process, the water is first in the gas phase and then enters the condensation process. Eventually, the water becomes liquid. According to the law of conservation of mass, the mass of water that was the initial vapor is equal to the mass of water that turns into a liquid. We can look at water chemistry. Conservation of mass can also be applied to chemical reactions. Let us consider oxygen and hydrogen. As a result of the reaction, water is formed. After the reaction, a new water product (H_2O) is formed. Oxygen (O_2) and hydrogen (H_2) are no longer present.

According to the law of conservation of mass, the sum of the masses of oxygen and hydrogen is equal to the mass of water. In general, according to the principle of conservation of mass, the total mass of the reactants in a chemical reaction must equal the total mass of the products.

A consequence of the law of conservation of mass is the continuity equation.

2.2.2. Continuity Equation in Cartesian Coordinates

Continuity equation based on the principles of conservation of mass, is written in Cartesian coordinates for the fluid passing through a fixed control volume seen in Figure 2.1.

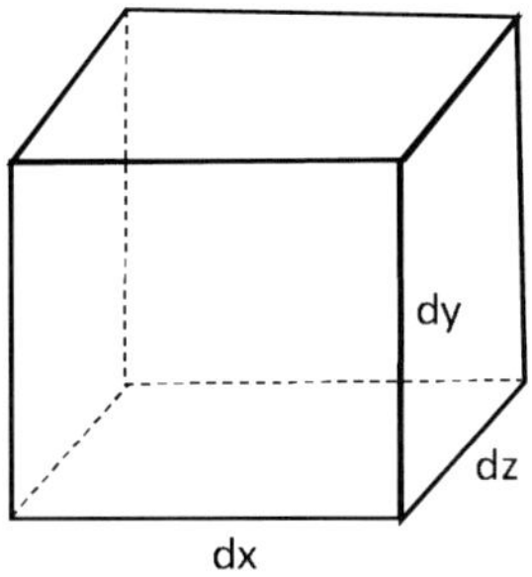

Figure 2.1. Conservation of mass in a control volume.

$$\frac{\partial \rho}{\partial t}+\frac{\partial(\rho u)}{\partial x}+\frac{\partial(\rho v)}{\partial y}+\frac{\partial(\rho w)}{\partial z}=s \tag{2.1}$$

where, u is the velocity component in x direction,
v is the velocity component in y direction,
w is the velocity component in z direction,
ρ is the fluid density,
t is the time.
S is the source.

For steady incompressible flow, ρ = constant and $\partial \rho/\partial t = 0$, and if there are no sources (s = 0), the preceding equation may be reduced to:

$$\frac{\partial u}{\partial x}+\frac{\partial v}{\partial y}+\frac{\partial w}{\partial z}=0 \tag{2.2}$$

For two dimensional flow, Equation (2.2) yields;

$$\frac{\partial u}{\partial x}+\frac{\partial v}{\partial y}=0 \tag{2.3}$$

For steady ($\partial \rho/\partial t = 0$) one dimensional flow, the continuity equation can be obtained from Equation (2.3). In Equation (2.3), for v = 0, we obtain $\partial u/\partial x = 0$, and u = constant, we can say that V = u. Therefore, for one dimensional flow, we can write "mass flow into control volume = mass flow out from control volume" which is mathematically expressed as,

$$\rho_1 A_1 V_1 = \rho_2 A_2 V_2 \quad (2.4)$$

where A, V are cross-section area and velocity of the flow, respectively.

For 1-D incompressible flow, if ρ = constant, we can write for a control volume (see Figure 2.2)

$$A_1 V_1 = A_2 V_2 = Q = \text{constant} \quad (2.5)$$

where Q is the flow rate.

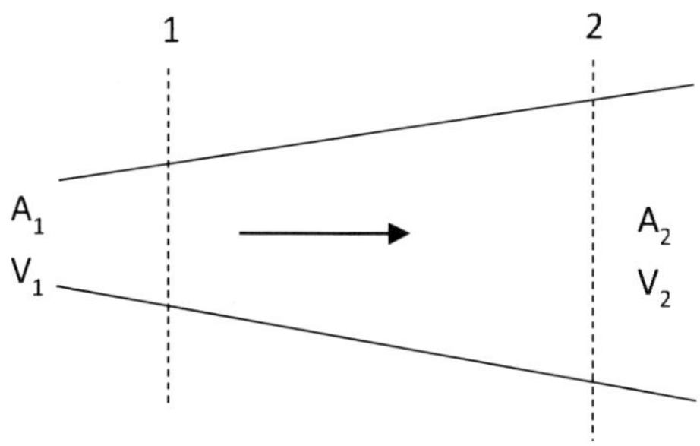

Figure 2.2. Control volume for 1-D incompressible flow.

2.2.3. Continuity Equation in Cylindrical Coordinates

The continuity equation is written as follows in the cylindrical coordinates shown in Figure 2.3,

$$\frac{\partial \rho}{\partial t} + \frac{\partial(\rho u)}{\partial r} + \frac{\rho u}{r} + \frac{1}{r}\frac{\partial(\rho v)}{\partial \theta} + \frac{\partial(\rho w)}{\partial z} = 0 \quad (2.6)$$

where, u is the velocity component in the radial r-direction,

v is the velocity component in the azimuthal θ-direction,

w is the velocity component in the z-direction.

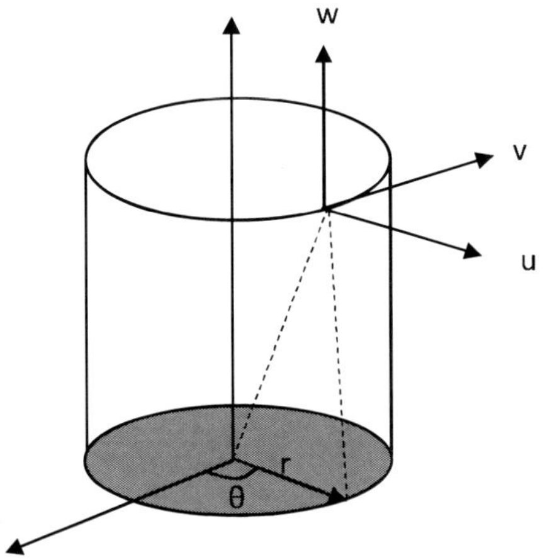

Figure 2.3. Velocity components in cylindrical coordinates.

For incompressible fluid flow, the preceding equation is reduced to:

$$\frac{\partial u}{\partial r} + \frac{u}{r} + \frac{1}{r}\frac{\partial v}{\partial \theta} + \frac{\partial w}{\partial z} = 0 \qquad (2.7)$$

For an axially symmetric flow, in which the axis r = 0 being the axis of symmetry, the term $(1/r)\partial v/\partial\theta = 0$, and Equation (2.7) reduces

$$\frac{\partial u}{\partial r} + \frac{u}{r} + \frac{\partial w}{\partial z} = 0 \qquad (2.8)$$

In cylindrical coordinates, for incompressible, plane, two-dimensional flow, Equation (2.7) reduces to (for w = 0),

$$\frac{\partial u}{\partial r} + \frac{u}{r} + \frac{1}{r}\frac{\partial v}{\partial \theta} = 0 \qquad (2.9)$$

2.3. Momentum Equation

2.3.1. General Remarks

Momentum (p) is the amount of motion of a moving object and is equal to its mass (m) times its velocity (v), (p = mv). According to the laws of physics, the force (F) and the time (t) product give momentum, (p = Ft). In other words, the section of momentum (p) over time (t) gives the force (F), ($F = p/t = mv/t = ma$).

Accordingly, the equation can be written in terms of force and mass as

$$F_{net} = ma \tag{2.10}$$

where F_{net} is the net force, m is the mass and a is the acceleration. This equation expresses mathematically the Newton's Second Law. According to this law, the change of momentum over time gives the force. For a fixed mass, force equals mass times acceleration. The equation given above is the mathematical form of Newton's Second Law (law of Motion).

Newton's Second Law is traditionally written as in Equation (2.10). In this equation, the net force applied to an object causes an acceleration that is inversely proportional to the mass of the object. Therefore, it is difficult to stop the objects with large masses or change their speed. Net force, F_{net}, is obtained by adding all the forces applied on the object. If there are 5 different forces acting on an object with mass m, the following equation can be written as,

$$F_{net} = \sum_{n=1}^{5} F_n = ma \tag{2.11}$$

or

$$F_{net} = F_g + F_P + F_\gamma + F_t + F_C \tag{2.12}$$

where, F_{net} is the net force,

F_g is the gravity force,

F_P is the pressure force,

F_γ is the viscosity force,

F_t is the turbulence force,

F_c is the compressibility force.

This is known as the equation of motion. When compressibility force F_C is negligible we can write as,

$$F_{net} = F_g + F_P + F_\gamma + F_t \tag{2.13}$$

This is called as the Reynold's equation of motion. When F_t is negligible,

$$F_{net} = F_g + F_P + F_\gamma \quad (2.14)$$

This is called the Navier- Stokes equation of motion. When $F_\gamma = 0$, we obtain

$$F_{net} = F_g + F_P \quad (2.15)$$

This is the ideal flow equation.

2.3.2. Momentum Equation in Cartesian Coordinates

Considering Equation (2.14) and applying Newton's Second Law to an incompressible fluid flow, the momentum equations are obtained in Cartesian coordinates as follows:

x-direction momentum equation

$$\rho(\frac{\partial u}{\partial t} + u\frac{\partial u}{\partial x} + v\frac{\partial u}{\partial y} + w\frac{\partial u}{\partial z}) = -\frac{\partial p}{\partial x} + \mu\left(\frac{\partial^2 u}{\partial x^2} + \frac{\partial^2 u}{\partial y^2} + \frac{\partial^2 u}{\partial z^2}\right) + \rho F_x \quad (2.16)$$

y-direction momentum equation

$$\rho(\frac{\partial v}{\partial t} + u\frac{\partial v}{\partial x} + v\frac{\partial v}{\partial y} + w\frac{\partial v}{\partial z}) = -\frac{\partial p}{\partial y} + \mu\left(\frac{\partial^2 v}{\partial x^2} + \frac{\partial^2 v}{\partial y^2} + \frac{\partial^2 v}{\partial z^2}\right) + \rho F_y \quad (2.17)$$

z-direction momentum equation

$$\rho\left(\frac{\partial w}{\partial t} + u\frac{\partial w}{\partial x} + v\frac{\partial w}{\partial y} + w\frac{\partial w}{\partial z}\right) = -\frac{\partial p}{\partial z} + \mu(\frac{\partial^2 w}{\partial x^2} + \frac{\partial^2 w}{\partial y^2} + \frac{\partial^2 w}{\partial z^2}) + \rho F_z \quad (2.18)$$

where, u, v, and w are the velocity components in x, y and z directions, respectively,

p is the pressure,

ρ is the density,

μ is the viscosity,

F_x, F_y and F_z are the external forces in x, y and z directions, respectively,

t is the time.

If variations in density are small, the density ρ may be replaced everywhere by the constant density ρ in the above equations,

$$\frac{\partial u}{\partial t}+u\frac{\partial u}{\partial x}+v\frac{\partial u}{\partial y}+w\frac{\partial u}{\partial z}=-\frac{1}{\rho}\frac{\partial p}{\partial x}+\nu\left(\frac{\partial^2 u}{\partial x^2}+\frac{\partial^2 u}{\partial y^2}+\frac{\partial^2 u}{\partial z^2}\right)+F_x \quad (2.19)$$

$$\frac{\partial v}{\partial t}+u\frac{\partial v}{\partial x}+v\frac{\partial v}{\partial y}+w\frac{\partial v}{\partial z}=-\frac{1}{\rho}\frac{\partial p}{\partial y}+\nu\left(\frac{\partial^2 v}{\partial x^2}+\frac{\partial^2 v}{\partial y^2}+\frac{\partial^2 v}{\partial z^2}\right)+F_y \quad (2.20)$$

$$\frac{\partial w}{\partial t}+u\frac{\partial w}{\partial x}+v\frac{\partial w}{\partial y}+w\frac{\partial w}{\partial z}=-\frac{1}{\rho}\frac{\partial p}{\partial z}+\nu\left(\frac{\partial^2 w}{\partial x^2}+\frac{\partial^2 w}{\partial y^2}+\frac{\partial^2 w}{\partial z^2}\right)+F_z \quad (2.21)$$

where, ν is the kinematic viscosity ($\nu = \mu/\rho$).

These are the general 3-D equations of incompressible fluid flow. These equations with together Equation (2.1) are known as Navier- Stokes equations.

For steady ($\partial\rho/\partial t = 0$) two-dimensional flow, the momentum equations can be written as,

$$u\frac{\partial u}{\partial x}+v\frac{\partial u}{\partial y}=-\frac{1}{\rho}\frac{\partial p}{\partial x}+\nu\left(\frac{\partial^2 u}{\partial x^2}+\frac{\partial^2 u}{\partial y^2}\right)+F_x \quad (2.22)$$

$$u\frac{\partial v}{\partial x}+v\frac{\partial v}{\partial y}=-\frac{1}{\rho}\frac{\partial p}{\partial y}+\nu\left(\frac{\partial^2 v}{\partial x^2}+\frac{\partial^2 v}{\partial y^2}\right)+F_y \quad (2.23)$$

2.3.3. Momentum Equation in Cylindrical Coordinates

The momentum equations can be written as follows in the cylindrical coordinates (r, θ, z),

r- component:

$$\frac{\partial u}{\partial t}+u\frac{\partial u}{\partial r}+\frac{v}{r}\frac{\partial u}{\partial \theta}+w\frac{\partial u}{\partial z}-\frac{v^2}{r}=-\frac{1}{\rho}\frac{\partial p}{\partial r}+\nu\left[\left(\frac{1}{r}\frac{\partial}{\partial r}\left(r\frac{\partial u}{\partial r}\right)+\frac{1}{r^2}\frac{\partial^2 u}{\partial \theta^2}-\frac{u}{r^2}-\frac{2}{r^2}\frac{\partial v}{\partial \theta}+\frac{\partial^2 u}{\partial z^2}\right]+F_r \quad (2.24)$$

θ- component:

$$\frac{\partial v}{\partial t}+u\frac{\partial v}{\partial r}+\frac{v}{r}\frac{\partial v}{\partial \theta}+w\frac{\partial v}{\partial z}-\frac{uv}{r}=-\frac{1}{\rho}\frac{\partial p}{\partial \theta}+\nu\left[\left(\frac{1}{r}\frac{\partial}{\partial r}\left(r\frac{\partial v}{\partial r}\right)+\frac{1}{r^2}\frac{\partial^2 v}{\partial \theta^2}-\frac{v}{r^2}+\frac{2}{r^2}\frac{\partial u}{\partial \theta}+\frac{\partial^2 v}{\partial z^2}\right]+F_\theta \quad (2.25)$$

z- component:

$$\frac{\partial w}{\partial t} + u\frac{\partial w}{\partial r} + \frac{v}{r}\frac{\partial w}{\partial \theta} + w\frac{\partial w}{\partial z} = -\frac{1}{\rho}\frac{\partial p}{\partial z} + \nu[(\frac{1}{r}\frac{\partial}{\partial r}(r\frac{\partial w}{\partial r}) + \frac{1}{r^2}\frac{\partial^2 w}{\partial \theta^2} + \frac{\partial^2 w}{\partial z^2}] - g + F_z \quad (2.26)$$

in these equations, u is the velocity component in the radial r-direction,
v is the velocity component in the azimuthal θ-direction,
w is the velocity component in the z-direction.

2.3.4. Bernoulli Equation

In the case of a steady flow of an inviscid fluid, the Bernoulli equation can be obtained from the momentum equations. The momentum equations under these conditions are:

x-component:

$$\frac{\partial u}{\partial t} + u\frac{\partial u}{\partial x} + v\frac{\partial u}{\partial y} + w\frac{\partial u}{\partial z} = -\frac{1}{\rho}\frac{\partial p}{\partial x} \quad (2.27)$$

y-component:

$$\frac{\partial v}{\partial t} + u\frac{\partial v}{\partial x} + v\frac{\partial v}{\partial y} + w\frac{\partial v}{\partial z} = -\frac{1}{\rho}\frac{\partial p}{\partial y} \quad (2.28)$$

z-component:

$$\frac{\partial w}{\partial t} + u\frac{\partial w}{\partial x} + v\frac{\partial w}{\partial y} + w\frac{\partial w}{\partial z} = -\frac{1}{\rho}\frac{\partial p}{\partial z} - g \quad (2.29)$$

Multiplying the first equation by u, the second by v, the third by w, and summing up the results yields:

$$\frac{\left(u\frac{\partial}{\partial x} + v\frac{\partial}{\partial y} + w\frac{\partial}{\partial z}\right)(u^2 + v^2 + w^2)}{2} = -u\left(\frac{\partial}{\partial x} + v\frac{\partial}{\partial y} + w\frac{\partial}{\partial z}\right)\left(\frac{p}{\rho}\right) - gw \quad (2.30)$$

Then, using (2.30) and noting that $w = (\partial u/\partial x + \partial v/\partial y + \partial w/\partial z)/2$, all terms can be taken to the left-hand side as:

$$(u\frac{\partial}{\partial x} + v\frac{\partial}{\partial y} + w\frac{\partial}{\partial z})\frac{u^2+v^2+w^2}{2} + \frac{p}{\rho} - gz = 0 \tag{2.31}$$

We note that the expression $(u^2 + v^2 + w^2)/2 + p/\rho + gz$ follows a simple equation, isomorphic to the density equation (2.30). Using $\gamma = g\rho$, we identify here the Bernoulli function given in (2.31) as

$$\frac{V^2}{2g} + \frac{p}{\gamma} + z = \text{Constant} \tag{2.32}$$

In left hand side of the equation, the first term is kinetic energy head, the second term is the pressure head and the third term is the potential head.

The Bernoulli equation is an equation of fluid dynamics and is derived from the Bernoulli principles. The Bernoulli principle, on the other hand, expresses that while the pressure energy and potential energy decrease in a flow, the flow speed increases simultaneously.

2.3.5. Momentum and Forces in Fluid Flow

In fluid mechanics, we can use the Newton's laws of equation. Applying Newton's second law to moving mass in the control volume yields,

$$\sum F = ma = \frac{d(mV)}{dt} \tag{2.33}$$

where, $\sum F$ is the sum of the external forces, m is the mass, a is the acceleration, V is the velocity, mV is the momentum, and t is the time.

If m is constant, we can write this equation as

$$\sum F\,dt = m dV \tag{2.34}$$

This equation is called as impulse-momentum equation in differential form which describes as "Impulse = Momentum." In difference form, we can write this equation as

$$F\Delta t = m\Delta V \tag{2.35}$$

where, $F\Delta t$ is impulse which is the product of the average net force acting on an object and its duration, $m\Delta V$ is the momentum which is the product of an object's mass and velocity change.

For formulation the acting force we consider a control volume as seen in Figure 2.4.

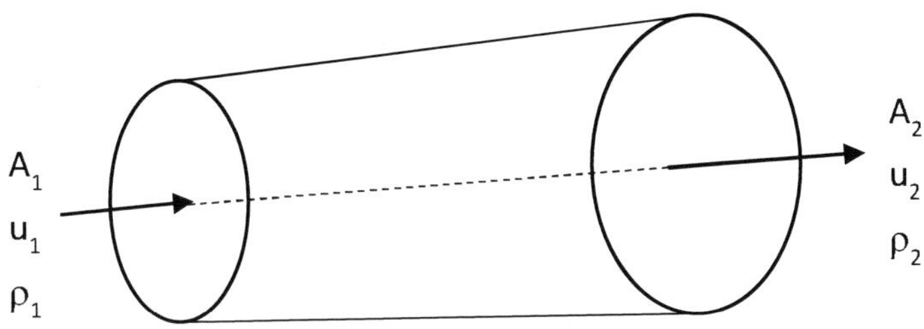

Figure 2.4. Control volume for acting force in fluid flow.

If the density of the fluid is constant, we can obtain the acting force due to flow from Equation (2.3) as

$$F = \rho Q(u_2 - u_1) \tag{2.36}$$

where, F is the acting force,

ρ is the fluid density,

Q is the rate of flow, $Q = m/\rho$,

u_1 and u_2 are the inflow and outflow velocities in control volume, respectively. As can be seen from Figure 2.5, acting forces in two-dimensional flow can be written as

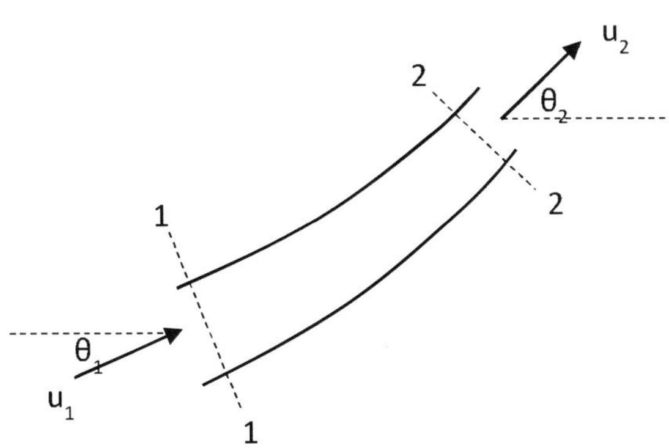

Figure 2.5. Two dimensional flow.

In x direction:

$$F_x = \rho Q(u_{2x} - u_{1x}) \tag{2.37}$$

In y direction:

$$F_y = \rho Q(u_{2y} - u_{1y}) \tag{2.38}$$

The resultant force can be shown as in Figure 2.6.

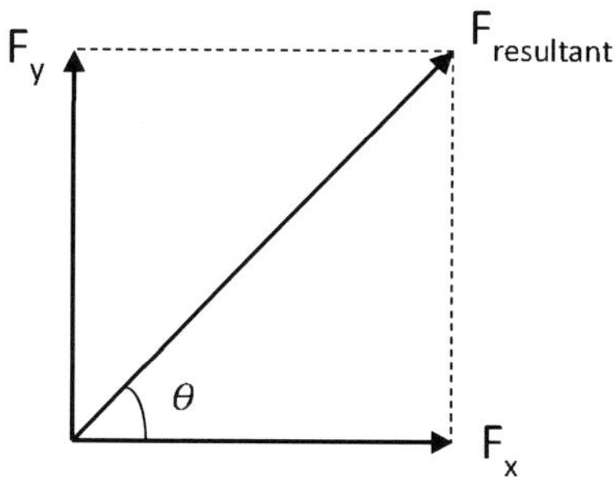

Figure 2.6. Resultant force diagram.

$$F = \sqrt{F_x^2 + F_y^2} \tag{2.39}$$

The angle of the force direction can be calculated as

$$\theta = \tan^{-1}(\frac{F_y}{F_x}) \tag{2.40}$$

2.4. Energy Equation

2.4.1. General Remarks

For a control volume shown in Figure 2.7, the total energy in the control volume is conserved. Conservation of energy can be expressed mathematically as

$$E_{in} - E_{out} = E_{st} + E_g \tag{2.41}$$

where, E_{in} is the inflow energy,
E_{st} is the stored energy,
E_g is the generated energy,
E_{out} is the outflow energy.

Figure 2.7. Control Volume for Energy Conservation.

Energy types are very diverse and can be classified in a variety of ways. Scientifically energy types are generally divided into two main groups. As seen from Figure 2.8, these are called Kinetic Energy and Potential Energy.

Kinetic energy is the energy resulting from the motion of bodies, waves, waters, atoms, molecules, and electrons.

Potential energy is location energy such as gravitational energy and stored energies. The sub-energy types of both types are shown in Figure 2.8.

In fluid dynamics, there are four types of energies as,

Kinetic energy (E_k),
Gravitational potential energy (U_g),
Elastic, spring potential energy (U_s), and
Heat (thermal) energy) (EH).

Using the principles of conservation of energy, we can write the following relationship between some initial time and at later time

$$E_{ki} + U_{gi} + U_{si} = E_{kf} + U_{gf} + U_{sf} + EH_f \tag{2.42}$$

where, Subscript i is the initial time,
Subscript f is the final time.
We can write Equation (2.42) in the following form:

$$\frac{1}{2}mv_i^2 + mgh_i + \frac{1}{2}kx_i^{\,2} = \frac{1}{2}mv_f^2 + mgh_f + \frac{1}{2}kx_f^2 + EH_f \tag{2.43}$$

where, m is the mass,
v is the velocity,
h is the potential head,
g is the gravitational acceleration,
x is the spring length,
k is the spring constant,
E_H is the heat (generated) energy.

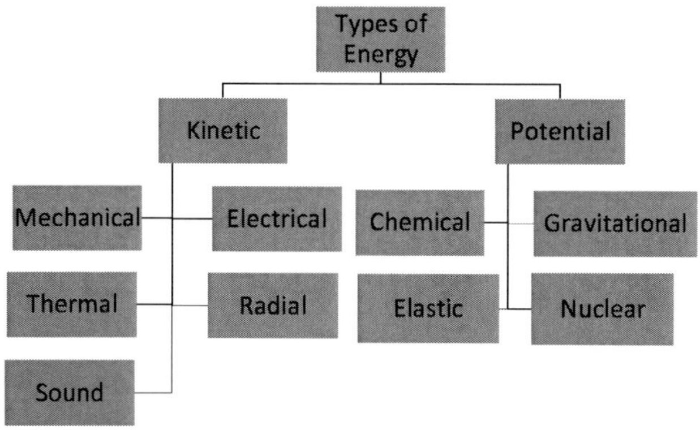

Figure 2.8. Types of energy.

2.4.2. Energy Equation in Cartesian Coordinates

For incompressible flow, the energy equation can be written in terms of static temperature:

$$\rho C_p \frac{\partial T}{\partial t} + \rho C_p u \frac{\partial T}{\partial x} + \rho C_p v \frac{\partial T}{\partial y} + \rho C_p w \frac{\partial T}{\partial z} = \frac{\partial}{\partial x}\left(k \frac{\partial T}{\partial x}\right) + \frac{\partial}{\partial y}\left(k \frac{\partial T}{\partial y}\right) + \frac{\partial}{\partial z}\left(k \frac{\partial T}{\partial z}\right) + q \qquad (2.44)$$

where, ρ is the density,
C_p is the specific heat at constant pressure,
T is the temperature,
k is the thermal conductivity,
u, v, and w are the velocity components in x, y, and z directions,
q is the local heat source per volume and per time.

Energy equation given in Equation (2.44) can be simplified by considering the fact that density is constant for incompressible flows as,

$$\rho C_p(\frac{\partial T}{\partial t} + u\frac{\partial T}{\partial x} + v\frac{\partial T}{\partial y} + w\frac{\partial T}{\partial z}) = k(\frac{\partial^2 T}{\partial x^2} + \frac{\partial^2 T}{\partial y^2} + \frac{\partial^2 T}{\partial z^2}) + q \quad (2.45)$$

In most engineering applications, energy equation reduces to a heat equation. In addition, the heat content of the fluid is mC_pT. So, we can write the following equation,

$$\frac{\partial(T)}{\partial t} + \frac{\partial(\rho uCpT)}{\partial x} + \frac{\partial(\rho vCpT)}{\partial y} + \frac{\partial(\rho wCpT)}{\partial z} = q \quad (2.46)$$

Then, taking the specific heat capacity C_p as constant, we obtain:

$$\frac{\partial T}{\partial t} + u\frac{\partial T}{\partial x} + v\frac{\partial T}{\partial y} + w\frac{\partial T}{\partial z} = q/(\rho C_p) \quad (2.47)$$

Finally, we can replace the actual density ρ by the constant value ρ_0 in the right-hand side.

There is linear relationship between temperature and density, and finally the energy equation for environmental fluids reduces to an equation governing density:

$$\frac{\partial \rho}{\partial t} + u\frac{\partial \rho}{\partial x} + v\frac{\partial \rho}{\partial y} + w\frac{\partial \rho}{\partial z} = \alpha q/C_p \quad (2.48)$$

where, q is the local heat source per volume and unit time.

If the internal heat sources are not present, the term q becomes zero. Equation (2.48) reduces to a statement as:

$$\frac{\partial \rho}{\partial t} + u\frac{\partial \rho}{\partial x} + v\frac{\partial \rho}{\partial y} + w\frac{\partial \rho}{\partial z} = 0 \quad (2.49)$$

2.4.3. Energy Equation in Cylindrical Coordinates

For constant thermal conductivity (k = constant), considering Equation (2.44), we can write similarly in cylindrical coordinates,

$$\rho Cp\left(\frac{\partial T}{\partial t} + v_r\frac{\partial T}{\partial r} + (v_\theta/r)\frac{\partial T}{\partial \theta} + (v_z\frac{\partial T}{\partial z}\right) = k[\frac{\partial^2 T}{\partial r^2} + \left(\frac{1}{r}\right)\frac{\partial T}{\partial r} + \left(\frac{1}{r^2}\right)\frac{\partial^2 T}{\partial \theta^2} + \frac{\partial^2 T}{\partial z^2}] + q \quad (2.50)$$

In cylindrical coordinates, we can write as similar equation as Equation (2.51):

$$\frac{\partial \rho}{\partial t} + u\frac{\partial \rho}{\partial r} + \frac{v}{r}\frac{\partial \rho}{\partial \theta} + w\frac{\partial \rho}{\partial z} = 0 \tag{2.51}$$

For steady state, two-dimensional flow

$$u\frac{\partial \rho}{\partial r} + \frac{v}{r}\frac{\partial \rho}{\partial \theta} = 0 \tag{2.52}$$

2.5. Transport Equation

2.5.1. General Remarks

The transport equation explains how a scalar quantity is transported in space. This equation is generally applied to transport quantities such as pollutant concentration, material properties, or temperature in air, water, or other objects. In mathematics, the transport equation is also called the convection-diffusion equation. This equation is a first order partial differential equation. The convection-diffusion equation forms the basis for many transport models (Stelling and Booij, 1999; Tey et al., 2017).

2.5.2. Convection-Diffusion Equation

The transport equation can be seen as the generalization form of the continuity equation. Let's consider a control volume as described in Figure 2.9.

Three-dimensional convection-diffusion equation can be written as:

$$\frac{\partial C}{\partial t} + \frac{\partial (uC)}{\partial x} + \frac{\partial (vC)}{\partial x} + \frac{\partial (wC)}{\partial z} = \frac{\partial}{\partial x}\left(D_x \frac{\partial C}{\partial x}\right) + \frac{\partial}{\partial y}\left(D_y \frac{\partial C}{\partial y}\right) + \frac{\partial}{\partial z}(D_z \frac{\partial C}{\partial z}) + S \tag{2.53}$$

where C is the scalar field to be analyzed, S is the source, u, v, and w denotes the convective velocities and the constant D_x, D_y, D_z denotes diffusion coefficients in the direction of x, y, and z direction, respectively.

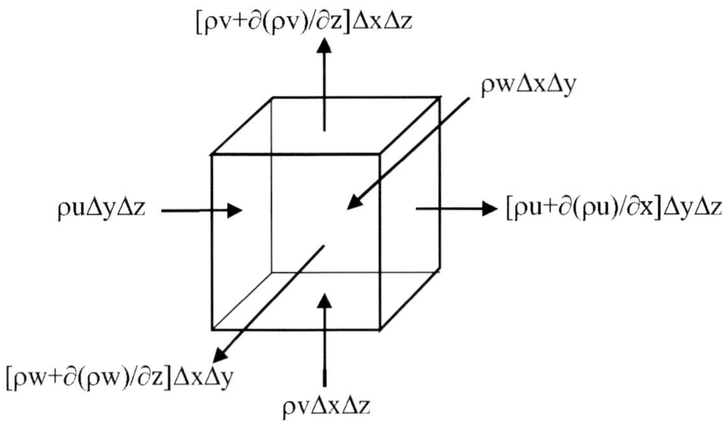

Figure 2.9. Control volume for Transport Equation.

In Equation (2.54), first term of the left hand side is the transient term, the other three terms are called convection terms, the first three terms of the lefts side are diffusion terms and S is the source term. Therefore, we can write as,

Transient term + Convection term = Diffusion term + Source term (2.54)

The transient term represents the accumulation of C in the control volume, the convection term represents the transport of C owing to the presence of the velocity field, the diffusion term represents the transport of C due to its gradients, and the source term S represents any source or sink.

In practice, both convection and diffusion are important problems in fluid mechanics. Convection is an important transport mechanism in many flows, but considering diffusion near boundaries gives more precise results in the solution. To determine which phenomenon is more effective in problem, we have to look at the dimensionless Peclet Number (see Subchapter 2.11.5).

If u, v, and w and D_x, D_y and D_z are not depend on x, y, z, the above equation reduces a simplified equation as,

$$\frac{\partial C}{\partial t} + u\frac{\partial C}{\partial x} + v\frac{\partial C}{\partial x} + w\frac{\partial C}{\partial z} = D\left(\frac{\partial^2 C}{\partial x^2} + \frac{\partial^2 C}{\partial y^2} + \frac{\partial^2 C}{\partial z^2}\right) + S \tag{2.55}$$

For two dimensional flow without source, we can write as

$$\frac{\partial C}{\partial t} + u\frac{\partial C}{\partial x} + v\frac{\partial C}{\partial x} = D(\frac{\partial^2 C}{\partial x^2} + \frac{\partial^2 C}{\partial y^2}) \quad (2.56)$$

This is the two-dimensional Convection-Diffusion equation. For one dimensional flow we can write it as,

$$\frac{\partial C}{\partial t} + u\frac{\partial C}{\partial x} = D\frac{\partial^2 C}{\partial x^2} \quad (2.57)$$

Equation (2.57) is known as the one dimensional form of the convection-diffusion equation or advection-diffusion equation.

2.5.3. Convection Equation

For three-dimensional flow, convection equation can be obtained from Equation (2.53) as,

$$\frac{\partial C}{\partial t} + u\frac{\partial C}{\partial x} + v\frac{\partial C}{\partial y} + w\frac{\partial C}{\partial z} = 0 \quad (2.58)$$

For two-dimensional flow, convection equation can be written as,

$$\frac{\partial C}{\partial t} + u\frac{\partial C}{\partial x} + v\frac{\partial C}{\partial y} = 0 \quad (2.59)$$

For one-dimensional flow, convection equation can be written as,

$$\frac{\partial C}{\partial t} + u\frac{\partial C}{\partial x} = 0 \quad (2.60)$$

Equation (2.60) is called a convection equation or advection equation.

2.5.4. Diffusion Equation

The diffusion equation can be obtained from Equation (2.55) as

$$\frac{\partial C}{\partial t} = D(\frac{\partial^2 C}{\partial x^2} + \frac{\partial^2 C}{\partial y^2} + \frac{\partial^2 C}{\partial z^2}) + S \quad (2.61)$$

This equation is referred as Fick's Second Diffusion law.
Special cases for diffusion equation:

For two-dimensional flow without source can be written as,

$$\frac{\partial C}{\partial t} = D\left(\frac{\partial^2 C}{\partial x^2} + \frac{\partial^2 C}{\partial y^2}\right) \tag{2.62}$$

For one- dimensional flow, the equation reduces to,

$$\frac{\partial C}{\partial t} = D\frac{\partial^2 C}{\partial x^2} \tag{2.63}$$

Equation (2.63) is called a diffusion equation.

2.5.5. Diffusion Equation in Cylindrical Coordinates

Three-dimensional diffusion equation in cylindrical coordinates can be written as

$$\frac{\partial C}{\partial t} = D\left(\frac{\partial^2 C}{\partial r^2} + \frac{1}{r}\frac{\partial C}{\partial r} + \frac{1}{r^2}\frac{\partial^2 C}{\partial \theta^2} + \frac{\partial^2 C}{\partial z^2}\right) + S \tag{2.64}$$

2.5.6. The Advection-Diffusion-Reaction Equation

We can write the advection-diffusion-reaction equation as,

$$\frac{\partial C}{\partial t} + u\frac{\partial C}{\partial x} = D\frac{\partial^2 C}{\partial x^2} + rC \tag{2.65}$$

in which C is the substance concentration, u is the transient velocity, D is the diffusion coefficient, x is the spatial variable, t is the temporal variable, r is the reaction factor.

2.6. Navier-Stokes Equations in Stream Function-Vorticity Form

2.6.1. General Remarks

The general flow equations referred to as the Navier-Stokes previously described are transformed into a new form by writing the stream function and the vorticity function instead of the velocity components such as u, v, w and p pressure, which are the primitive variables. Thus, a new set of partial differential equations is obtained. In these new equations, while the number of unknowns decreases, the derivative order of the equations increases.

2.6.2. Vorticity – Stream Function in Cartesian Coordinates

Consider the two-dimensional incompressible flow. The governing equations of two-dimensional unsteady incompressible viscous flow are the continuity equation, and the two components of the momentum equations. We can obtain the vorticity-stream function equations in Cartesian coordinates using these equations.

First of all, pressure term P is removed from momentum equations to obtain the stream function-vorticity ($\psi - \xi$) equation. The definition of vorticity is known as:

$$\xi = \frac{\partial v}{\partial x} - \frac{\partial u}{\partial y} \tag{2.66}$$

where, ξ is the vorticity,

u, and v, are the velocity components along x and y axis, respectively.

Using the two dimensional momentum equations and considering Equation (2.66) we can obtain

$$\frac{\partial \xi}{\partial t} + u\frac{\partial \xi}{\partial x} + v\frac{\partial \xi}{\partial y} = \frac{1}{Re}\left(\frac{\partial^2 \xi}{\partial x^2} + \frac{\partial^2 \xi}{\partial y^2}\right) \tag{2.67}$$

where R_e is the Reynolds number.

Equation (2.67) is a parabolic type partial differential equation called vorticity-transport equation. This equation is also referred as convection-diffusion equation.

Defining a stream function ψ yields as

$$\frac{\partial \psi}{\partial x} = -v \tag{2.68}$$

$$\frac{\partial \psi}{\partial y} = u \tag{2.69}$$

ψ identically, satisfies the continuity equation of two dimensional flow. Further, use of definition of ξ leads to

$$\frac{\partial^2 \psi}{\partial x^2} + \frac{\partial^2 \psi}{\partial y^2} = -\xi \tag{2.70}$$

This is an elliptic partial differential equation called Poisson equation. Substituting Equation (2.70) into Equation (2.67) yields,

$$\frac{\partial \xi}{\partial t} + \frac{\partial \psi}{\partial y}\frac{\partial \xi}{\partial x} + \frac{\partial \psi}{\partial x}\frac{\partial \xi}{\partial y} = \frac{1}{R_e}\left(\frac{\partial^2 \xi}{\partial x^2} + \frac{\partial^2 \xi}{\partial y^2}\right) \tag{2.71}$$

Equation (2.70) and Equation (2.71) are known as the Navier-Stokes equations in vorticity-stream function form. The former one called as Poisson equation, the latter one called also Advection –Diffusion equation.

2.6.3. Stream Function for Plane Two-Dimensional Flow in Cartesian Coordinates

Consider the schematics of plane flows in Cartesian coordinates with velocity components as seen in Figure 2.10 as u(x, y), v(x, y), and w = 0.

Using Equation (2.68), Equation (2.69), and Equation (2.70), vorticity equation can be obtained as:

$$\frac{\partial \xi_z}{\partial t} + \frac{\partial \psi}{\partial y}\frac{\partial \xi_z}{\partial x} - \frac{\partial \psi}{\partial x}\frac{\partial \xi_z}{\partial y} = -\nu\left(\frac{\partial^4 \psi}{\partial x^4} + \frac{\partial^4 \psi}{\partial x^2 \partial y^2} + \frac{\partial^4 \psi}{\partial y^4}\right) \tag{2.72}$$

We can find the flow rate passing through two adjacent streamlines: $Q = \psi_2 - \psi_1$.

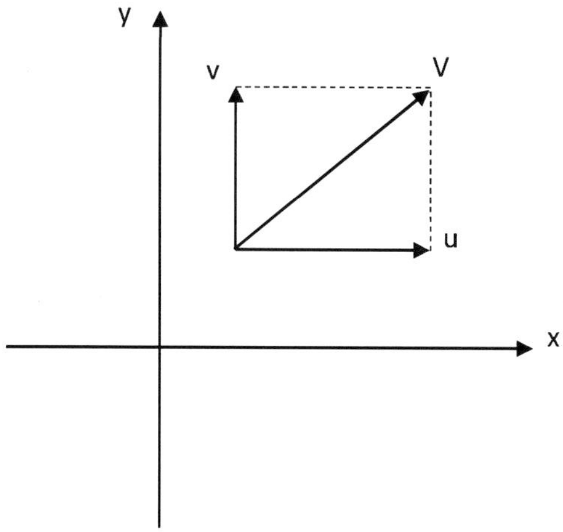

Figure 2.10. Schematics of plane flows in a Cartesian coordinate system.

2.6.4. Vorticity – Stream Function in Cylindrical Coordinates

We can derive the equations used to solve flow in a cylindrical coordinate in axially symmetric flow as can be seen in Figure 2.11 with velocity components as $v_\Theta = 0$, v_z, and v_r.

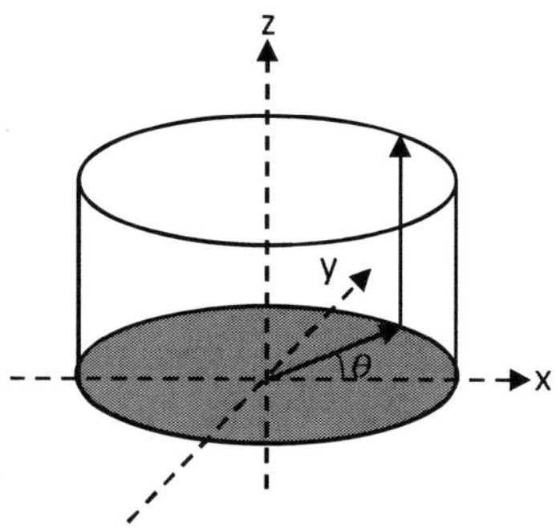

Figure 2.11. Cylindrical coordinate system.

Using the Stokes stream function ψ, velocity components of an incompressible axisymmetric flow can be characterized as

$$u_z = u = \frac{1}{r}\frac{\partial \psi}{\partial r} \quad (2.73)$$

$$u_r = v = -\frac{1}{r}\frac{\partial\psi}{\partial z} \qquad (2.74)$$

The azimuthal component u_θ does not depend on the stream function and can be defined independently. In this writeup, $u_\theta = 0$. Also, this formulation automatically satisfies continuity, since

$$\frac{1}{r}\frac{\partial(ur)}{\partial r} + \frac{\partial(uz)}{\partial z} = -\frac{1}{r}\frac{\partial^2\psi}{\partial r\,\partial z} + \frac{1}{r}\frac{\partial^2\psi}{\partial z\,\partial r} \qquad (2.75)$$

When $\partial/\partial\theta = 0$ and $u_\theta = 0$ in an axisymmetric flow, only the w_θ component remains, and we obtain vorticity.

$$\xi_\theta = \frac{\partial v}{\partial z} - \frac{\partial u}{\partial r} \qquad (2.76)$$

Definitions of velocity are replaced by those based on the Stokes stream function, and the result is:

$$-\xi_r = \frac{\partial^2\psi}{\partial z^2} + \frac{\partial^2\psi}{\partial r^2} - \frac{1}{r}\frac{\partial\psi}{\partial r} \qquad (2.77)$$

The vorticity transfer equation provides information about the temporal evolution of vorticity. Typically, the momentum equation's curl is used to derive this equation.

For the axisymmetric flow, we have

$$\frac{\partial\xi}{\partial t} + u\frac{\partial w}{\partial z} + v\frac{\partial v}{\partial r} = \nu\left(\frac{\partial^2\xi}{\partial z^2} + \frac{\partial^2\xi}{\partial r^2} + \frac{1}{r}\frac{\partial\xi}{\partial r}\right) \qquad (2.78)$$

The flow rate contained by the streamtube can be calculated using $Q = 2\pi\psi$ for axisymmetric flow.

2.6.5. Stream Function for Plane Two-Dimensional Flow in Cylindrical Coordinates

We can consider the stream function for plane two-dimensional flow in cylindrical coordinates (r, θ, z) as shown in Figure 2.12, with velocities as v_r, u_θ, $v_z = 0$.

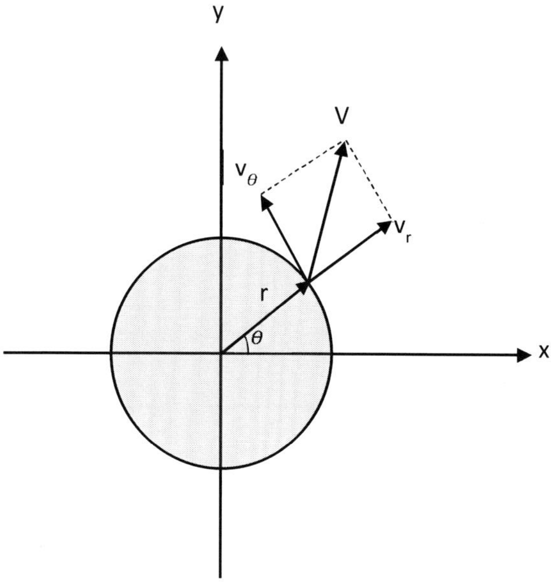

Figure 2.12. Schematics of plane flows in a polar coordinate system.

Definition of stream function f = ψ(r, θ) as,

$$v_r = \frac{1}{r}\frac{\partial \psi}{\partial r} \quad (2.79)$$

$$v_\theta = -\frac{\partial \psi}{\partial \theta} \quad (2.80)$$

$$-\xi_z = \frac{\partial^2 \psi}{\partial r^2} + \frac{1}{r}\frac{\partial \psi}{\partial r} + \frac{\partial^2 \psi}{\partial \theta^2} = \frac{1}{r}\frac{\partial \psi}{\partial r} + \frac{\partial^2 \psi}{\partial \theta^2} = \frac{1}{r}\frac{\partial}{\partial r}(r\frac{\partial \psi}{\partial r}) + \frac{1}{r^2}\frac{\partial^2 \psi}{\partial \theta^2} \quad (2.81)$$

Vorticity equation is given as:

$$\frac{\partial \xi_z}{\partial t} + v_r\frac{\partial \xi_z}{\partial r} + v_\theta\frac{1}{r}\frac{\partial \xi_z}{\partial \theta} = \nu\frac{1}{r}\frac{\partial}{\partial r}(r\frac{\partial \xi_z}{\partial r}) + \nu\frac{1}{r^2}\frac{\partial^2 \xi_z}{\partial \theta^2} \quad (2.82)$$

We can find the flow rate between streamlines as: $Q = \psi_2 - \psi_1$.

2.7. Saint-Venant Equations

2.7.1. General Remarks

The one-dimensional Saint-Venant equations or their simplified form are often used to solve flow problems in open channels. In these equations, the variation of flow rate, velocity, and water depth over time and space is studied. The Saint-Venant equations consist of the continuity equation and the momentum equation. These equations are partial differential equations in terms of space and time.

Theoretically, flow in streams is actually three-dimensional: longitudinal, transverse, and lateral. Moreover, the flow changes with time. The equations of the three-dimensional and time-dependent flow system can be derived from the Navier-Stokes equations. But the resulting equations are very complicated. To solve them, a lot of data is needed. Therefore, the problems are studied in one-dimensional and time-dependent form by considering only the longitudinal dimension, which is more effective than the lateral and transverse directional changes of the flow. The basic equations of one-dimensional and time-dependent flow in open channels were derived by Barre de Saint-Venant in the 1870s.

2.7.2. Continuity Equation

The general principle of the continuity equation can be written as,

Rate of change of mass = Mass inflow - Mass outflow.

Considering Figure 2.13, the continuity equation for unsteady flow in open cannels with any irregular cross section can be written as

$$\frac{\partial A}{\partial t} + \frac{\partial Q}{\partial x} - q = 0 \tag{2.83}$$

where, A is the cross sectional area,

Q is the rate of flow,

q is the lateral inflow,

x is the spatial coordinate,

t is the time.

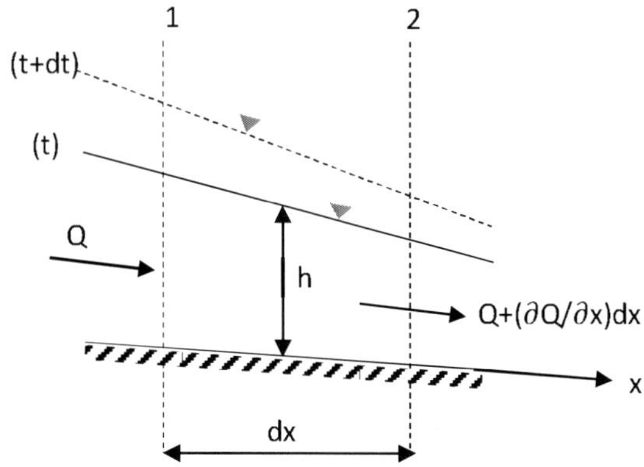

Figure 2.13. Configuration of Open Channel flow for Saint-Venant Equations.

2.7.3. Momentum Equation

The momentum equation of one-dimensional unsteady flow in open channels can be expressed as

$$\frac{1}{A}\frac{\partial Q}{\partial t} + \frac{1}{A}\frac{\partial}{\partial x}\left(\frac{Q^2}{A}\right) + g\frac{\partial h}{\partial x} - g(S_o - S_f) = 0 \qquad (2.84)$$

where, A is the cross sectional area,

Q is the discharge,

h is the water depth,

S_o is the bottom slope,

S_f is the friction slope,

G is the gravitational acceleration.

In Equation (2.84); the First term is the local acceleration term, the second term is convective acceleration term, the third is the pressure force term, the forth term is the gravity force term, and the fifth term is friction force term.

We can write the momentum equation in another form as

$$\frac{\partial V}{\partial t} + V\frac{\partial V}{\partial x} + g\frac{\partial h}{\partial x} - g(S_o - S_f) = 0 \qquad (2.85)$$

where, V is the flow velocity.

As a summary, Equation (2.83) with Equation (2.84) or Equation (2.85) is referred as Saint-Venant equations which are valid for open channel flows.

Depending on the analyzed problem, some simplifications can be done in momentum equation.

In momentum equation, Equation (2.85);

If there is no neglected term, the resultant equation with continuity equation is referred to Dynamic Wave Model.

If the first two terms are neglected, the resultant equation with continuity equation is referred to Diffusion Wave Model.

If the first three terms are neglected, the resultant equation with continuity equation is referred to Kinematic Wave Model.

2.7.4. Saint-Venant Equations for Steady State Flow

For steady state case, Saint-Venant equations, Equation (2.83) and Equation (2.84), simplify to:

$$\frac{dQ}{dx} - q = 0 \tag{2.86}$$

$$\frac{1}{A}\frac{\partial}{\partial x}\left(\frac{Q2}{A}\right) + g\frac{\partial h}{\partial x} - g(S_o - S_f) = 0 \tag{2.87}$$

2.8. Vorticity

Vorticity can be defined as the rotation of fluid elements. Vorticity is a vector, and its unit is 1/second. The three components of the vorticity can be written as,

$$\xi_x = \frac{\partial w}{\partial y} - \frac{\partial v}{\partial z} \tag{2.88}$$

$$\xi_y = \frac{\partial u}{\partial z} - \frac{\partial w}{\partial x} \tag{2.89}$$

$$\xi_z = \frac{\partial v}{\partial x} - \frac{\partial u}{\partial y} \tag{2.90}$$

where, ξ is the vorticity,

u, v, and w are the velocity components in x, y and z directions, respectively.

For a two dimensional flow the vector is always normal to the flow field plane. Vorticity magnitude is calculated using the following formula,

$$\xi = \sqrt{\xi_x^2 + \xi_y^2 + \xi_z^2} \tag{2.91}$$

If the frictional force is not present, the equations are written in terms of the individual vorticity components as:

$$\frac{d\xi_x}{dt} = \xi_x \frac{\partial u}{\partial x} + \xi_y \frac{\partial u}{\partial y} + \xi_z \frac{\partial u}{\partial z} - \frac{g}{\rho_0} \frac{\partial p}{\partial y} \tag{2.92}$$

$$\frac{d\xi_y}{dt} = \xi_x \frac{\partial v}{\partial x} + \xi_y \frac{\partial v}{\partial y} + \xi_z \frac{\partial u}{\partial z} + \frac{g}{\rho_0} \frac{\partial p}{\partial x} \tag{2.93}$$

$$\frac{d\xi_z}{dt} = \xi_x \frac{\partial w}{\partial x} + \xi_y \frac{\partial w}{\partial y} + \xi_z \frac{\partial w}{\partial z} - \frac{g}{\rho_0} \frac{\partial p}{\partial y} \tag{2.94}$$

If the formulation includes Coriolis terms, the Coriolis parameter is added to the vertical vorticity component ξ_z:

$$\xi_z \rightarrow \xi_z + f \tag{2.95}$$

This means that the vertical vorticity component is added by vorticity due to the ambient rotation (see in Subchapter 2.10.2).

For rotational fluid vertical vorticity component should be written as

$$\xi_z = \frac{1}{2}\left(\frac{\partial v}{\partial x} - \frac{\partial u}{\partial y}\right) \tag{2.96}$$

2.9. Potential Flow

Potential flow or irrotational flow refers to a flow outside of the boundary layer zone in which potential flow principles are valid. In this flow we can neglect the viscose effect in the flow domain.

2.9.1. Velocity Potential Function

Potential function is used in the potential flow. Assuming an incompressible, viscid and irrotational flow, the velocity potential is a continuous function that satisfies the basic laws of fluid dynamics such as conservation of mass and conservation of momentum.

The components of velocity in Cartesian coordinates can be written as,

$$u = \frac{\partial \phi}{\partial x} \tag{2.97}$$

$$v = \frac{\partial \phi}{\partial y} \tag{2.98}$$

$$w = \frac{\partial \phi}{\partial z} \tag{2.99}$$

where ϕ is the velocity potential function,

u, v, and w are velocity components x, y, and z directions, respectively.

2.9.2. Stream Function

Stream function can be defined as a function for two dimensional flow or three dimensional axially symmetric flows. We define the stream function as a function of velocity components in Equation (2.68), and Equation (2.69).

The relationship between stream function, ψ, and velocity potential function, ϕ, can be written as,

$$\frac{\partial \phi}{\partial x} = \frac{\partial \psi}{\partial y} \tag{2.100}$$

$$\frac{\partial \phi}{\partial y} = -\frac{\partial \psi}{\partial x} \tag{2.101}$$

2.9.3. Laplace Equation in Cartesian Coordinates

The flow must still satisfy the continuity equation. Considering the continuity equation as seen in Equation (2.2), we can write

$$\frac{\partial^2 \phi}{\partial x^2} + \frac{\partial^2 \phi}{\partial y^2} + \frac{\partial^2 \phi}{\partial z^2} = 0 \quad (2.102)$$

This equation is referred to Laplace equation in Cartesian coordinates.

2.9.4. Laplace Equation in Cylindrical Coordinates

Using the geometric relationship, we can write in cylindrical coordinates

$$r^2 = x^2 + y^2 \quad (2.103)$$

and

$$\vartheta = \tan^{-1}\left(\frac{y}{x}\right) \quad (2.104)$$

The Laplace equation in cylindrical coordinates can be written as

$$\frac{\partial^2 \phi}{\partial r^2} + \frac{1}{r}\frac{\partial \phi}{\partial r} + \frac{1}{r^2}\frac{\partial^2 \phi}{\partial \theta^2} + \frac{\partial^2 \phi}{\partial z^2} = 0 \quad (2.105)$$

2.9.5. Potential Lines and Streamlines

The lines of constant potential function, ϕ, are called potential lines, or equipotential lines, of the flow. For equipotential line, ϕ = Constant. In two-dimensional flow, we can write as

$$d\phi = \frac{\partial \phi}{\partial x}dx + \frac{\partial \phi}{\partial y}dy \quad (2.106)$$

$$d\phi = u dx + v dy \quad (2.107)$$

Since $d\phi = 0$ along a potential line, we have

$$\frac{dy}{dx} = -\frac{u}{v} \quad (2.108)$$

This equation refers to potential lines. Here, dy/dx is the slope of equipotential line.

A streamline is a line everywhere tangent to the velocity at a given instant. For streamline, Ψ = Constant. Similarly, for stream function we can write $d\Psi = (\partial\Psi/\partial x)dx + (\partial\Psi/\partial y)dy = 0$. The streamline equation is given as,

$$\frac{dy}{dx} = \frac{v}{u} \tag{2.109}$$

Here, dy/dx is the slope of streamline. There is a principle that equipotential lines are always perpendicular to the streamlines for a inviscid and irrotational flow. Streamlines with equipotential lines together describes the flownet as can be seen in Figure 2.14.

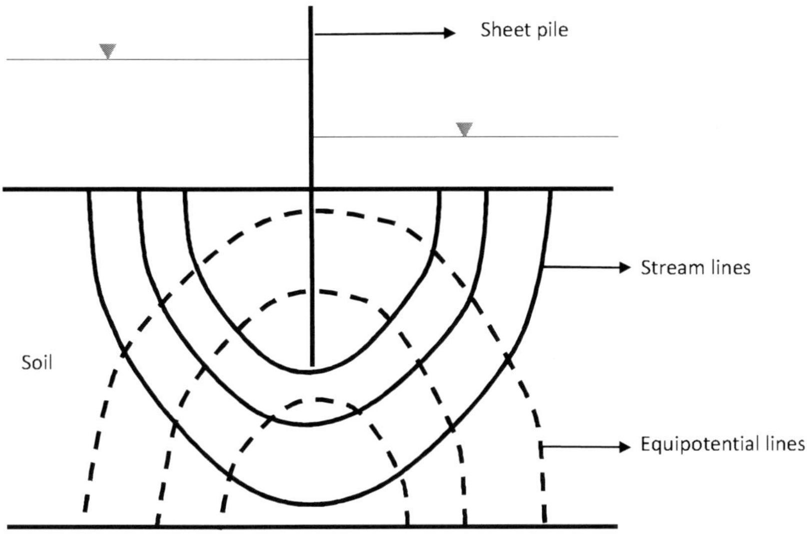

Figure 2.14. An illustration for flow net consists of streamlines and equipotential lines.

2.10. Hydrostatic Balance and Earth's Rotation

2.10.1. Hydrostatic Balance

Hydrostatic balance can be derived by Navier- Stokes equations. For the equilibrium situation we can write for u = 0, v = 0, $\frac{\partial p}{\partial x} = 0$, and $\frac{\partial p}{\partial y} =$

0, where u and v are the velocity components in x, and y directions, respectively, p is the pressure.

Substituting these equations into Navier- Stokes equations yields

$$\frac{\partial p}{\partial z} + \rho g = 0 \tag{2.110}$$

This is the hydrostatic approximation equation or hydrostatic balance equation.

Solving this equation gives:

$$p = \gamma z \tag{2.111}$$

This expression refers to pressure in the case of a static fluid. As can be seen, in a static fluid, pressure depends only on fluid depth and the pressure acting on a point is the same in all directions.

2.10.2. Earth's Rotation

If the horizontal dimension of the flow is large and the velocity components are small, the rotation effects will be important due to the rotation of the earth. In this situation, Coriolis terms must be added to the momentum equations. Coriolis parameter can be written as:

$$f = 2\Omega \sin\Theta \tag{2.112}$$

here, Θ is the latitude,

Ω is the angular rate of rotation of the earth.

Applying hydrostatic balance for vertical component, the momentum equations become:

$$\frac{\partial u}{\partial t} + u\frac{\partial u}{\partial x} + v\frac{\partial u}{\partial y} + w\frac{\partial u}{\partial z} - f_v = -\frac{1}{\rho}\frac{\partial p}{\partial x} + F_x \tag{2.113}$$

$$\frac{\partial v}{\partial t} + u\frac{\partial v}{\partial x} + v\frac{\partial v}{\partial y} + w\frac{\partial v}{\partial z} - f_u = -\frac{1}{\rho}\frac{\partial p}{\partial y} + F_y \tag{2.114}$$

$$\frac{\partial p}{\partial z} = -\rho g \tag{2.115}$$

where f_u and f_v are the Coriolis parameters in x and y directions, respectively.

2.11. Scales and Dimensionless Numbers

2.11.1. Definition of Scales

Solving the most general forms of the Navier-Stokes equations is very difficult and unrealistic. Some approaches applied in specific situations are much more useful. To proceed systematically, one needs a procedure to help understand exactly which term is small and which is not. A standard procedure would be to first find the scales that are related to the problem being addressed. Depending on the size of these scales, dimensionless parameters can be found that indicate the relative importance of the different parts of complete equations.

For a problem in hydraulics or fluid mechanics, there is the length scale L, the time scale T that has elapsed since the problem began, and the velocity scale U for the speed of the problem. If gravity is expected to play a role in the problem, the gravitational acceleration g can be used as a measure of body force per unit mass.

Dimensionless parameters commonly used in hydraulics are the Froude number, the Richardson number, the Reynolds number, the Peclet number, and the Rossby number. We have discussed these parameters in the following sections.

2.11.2. The Froude Number

One of the dimensionless numbers used in hydraulics is Froude number. The Froude number can be obtained from general Momentum equations. The ratio of inertia and pressure force can be written as inertia/ pressure force = $(u^2/L)/(gh/L) = u^2/gh$.

The Froude number is defined as the square root of this ratio:

$$F_r = \frac{u}{\sqrt{gh}} \tag{2.116}$$

where, F_r is the Froude number,

u is the velocity,
g is the gravitational acceleration,
h is the characteristic depth.

Froude number can be also interpreted as the ratio of the velocity and wave celerity. In hydraulics, the following flow regimes are distinguished:

If $F_r < 1$, the flow is subcritical,
If $F_r = 1$, the flow is critical,
If $F_r > 1$, the flow is supercritical.

2.11.3. The Richardson Number

In the determination of effect of buoyancy on stability of a body, the Richardson number can be used. Richardson number compares buoyancy forces to inertia, or, alternatively, potential energy to kinetic energy.

Richardson number can be defined as the ratio of pressure force to inertial force:

$$R_i = \frac{gh\Delta\rho}{\rho u^2} \qquad (2.117)$$

where, R_i is the Richardson number,
h is the fluid depth,
ρ is the fluid density,
u is the velocity,
g is the gravitational acceleration.

The Richardson number shows the ratio of potential energy over kinetic energy. The physical interpretation of the Richardson number can be considered as:

If $R_i \ll 1$, stratification is weak and vulnerable to mixing,
If $R_i \sim 1$, stratification is important and may be reduced by mixing,
If $R_i \gg 1$, stratification is strong and resisting mixing.

2.11.4. The Reynolds Number

The Reynolds number is used to show the ratio of inertia force to friction force in a flow. The friction force is caused by viscosity. Therefore, the Reynolds number, R_e, is defined as the ratio of inertia to viscous force as $R_e = (u^2/L)/[(\mu u)/\rho L^2)] = \rho uL/\mu$. So, Reynolds number is defined as,

$$R_e = \frac{\rho uL}{\mu} \tag{2.118}$$

where, R_e is the Reynolds number,
ρ is the density,
u is the velocity,
L is the characteristic length,
μ is the viscosity of fluid.
For a pipe flow, L indicates diameter of the pipe as D.

Reynolds number is a key in distinguishing between laminar and turbulent flows.

Using Reynolds number, R_e, we can classify the pipe flow as

If $R_e < 2100$, the pipe flow is laminar,
If $2100 < R_e < 4000$, the pipe flow is critical,
If $R_e > 4000$, the pipe flow is turbulent.

2.11.5. The Peclet Number

Another number is the Peclet number. It is defined as,

$$P_e = \frac{uL}{D} \tag{2.119}$$

where, u is the velocity,
L is the length,
D is the diffusivity of the fluid.

The Peclet number indicates the ratio of advection over dispersion in transporting the substance across the system.

If $P_e \ll 1$, advection is much weaker than diffusion.

If $P_e \gg 1$, advection is much stronger than diffusion.

If $P_e \sim 1$, advection and diffusion are equally important, and neither may be ignored.

2.11.6. The Rossby Number

One another dimensionless number should be considered if the Coriolis terms are retained in the momentum equations to consider the effects of the earth's rotation.

According to Navier- stokes equations without Coriolis effects, the additional terms are −fv and +fu, respectively, where $f = 2\Omega \sin\Theta$. The ratio inertia/Coriolis yields as = $(u^2/L)/\Omega u = u/\Omega L$.

This dimensionless parameter is called the Rossby number:

$$R_0 = \frac{u}{\Omega L} \tag{2.120}$$

where R_o is the Rossby number,

Ω is the earth's rotation rate,

u is the velocity,

L is the length.

The meaning of the Rossby number can be used as follows:

If $R_o \ll 1$ or $R_o \sim 1$, the earth's rotation has important effects and must be retained in the formulation of the problem, but

If $R_o \gg 1$, the earth's rotation is not important and can be ignored.

Chapter 3

Some Analytical Solutions

3.1. Introduction

As discussed in Chapter 2, differential equations are common in the study of many events in science and engineering. Precise numerical approximate solutions can be obtained over certain ranges using methods such as finite differences and finite elements in the solution of differential equations. Moreover, analytical solutions of high order differential equations or strong nonlinear differential equations are still difficult to obtain. However, in the field of machine learning, the analytical solution can be considered as a research analog and used as training data with arbitrary initial and boundary condition equations. To this end, this section explains some analytical solutions that are used in the solution of simplified ordinary or partial differential equations and can be used in Microsoft Excel in fluid mechanics (hydraulics) problems.

In general, the differential equations for fluid flow are second order partial differential equations (Meleshko, 2005; Chasnov, 2019). However, under certain conditions, the partial differential equations can be simplified to first order partial differential equations or even to ordinary differential equations as mentioned in Chapter 2.

3.2. Sample Equations

Differential equations can be classified in different ways. They have three basic types: 1. ordinary, 2. partial, and 3. differential-algebraic They can be also classified by attributes such as their order, linearity, and degree. We are interested in differential equation used in the problems of hydraulics and fluid mechanics.

3.2.1. Some Differential Equations in Hydraulics

The following differential equations that derived from general governing equations are used commonly in hydraulics and fluid mechanics:

The simple water balance equation

$$I - O = \frac{dS}{dt} \tag{3.1}$$

here, I is the inflow rate,

O is the outflow rate and

S is the storage volume.

Equation (3.1) is an ordinary differential equation.

The continuity equation for one-dimensional unsteady gradually-varied flow

$$\frac{\partial A}{\partial t} + \frac{\partial Q}{\partial x} = 0 \tag{3.2}$$

where, A is the cross-sectional area,

Q is the flow rate,

x and t are the space and time coordinates, respectively.

Equation (3.2) is a partial differential equation.

General form of the Gradually Varied Flow equation is:

$$\frac{dy}{dx} = \frac{S_0 - S_f}{1 - Fr^2} \tag{3.3}$$

where: S_o is the bottom slope,

S_f is the friction slope,

y is the water depth,

x is the longitudinal distance,

Fr is the Froude number.

Equation (3.3) is an ordinary differential equation.

The initial value problems for the differential equations that only contain derivatives with regard to time t are used to characterize the water quality of lakes as follows:

$$\frac{d\varphi}{dt} + \frac{\varphi}{T} = \frac{\varphi_e}{T} \tag{3.4}$$

where, $\varphi(t)$ is the Pollutant concentration at any given time,

φ_e is the equilibrium concentration,

T is the relaxation time.

The one-dimensional wave equation can be written as

$$\frac{\partial^2 \phi}{\partial t^2} - c^2 \frac{\partial^2 \phi}{\partial x^2} = 0 \tag{3.5}$$

where c is the expected constant propagation speed. The propagation of oscillations at a fixed speed in some quantity, ϕ is described by Equation (3.5), a linear second-order partial differential equation.

Two-dimensional steady compressible potential flow around a slim body can be given as

$$(1 - M^2)\frac{\partial^2 \phi}{\partial x^2} + \frac{\partial^2 \phi}{\partial y^2} = 0 \tag{3.6}$$

where M is the Mach number that is constant. The Mach number takes values less than 1 or greater than 1.

The Laplace equation

$$\frac{\partial^2 \phi}{\partial x^2} + \frac{\partial^2 \phi}{\partial y^2} = 0 \tag{3.7}$$

where ϕ is the stream function or potential function. Equation (3.7) is a second-order partial differential equation.

The one-dimensional convection-diffusion equation

$$\frac{\partial \phi}{\partial t} + c\frac{\partial \phi}{\partial x} = D\frac{\partial^2 \phi}{\partial x^2} \tag{3.8}$$

where D is the diffusion coefficient and c is the propagation speed, c and D are taken to be constants.

The diffusion or heat conduction equation

$$\frac{\partial \phi}{\partial t} = D\frac{\partial^2 \phi}{\partial x^2} \tag{3.9}$$

which is unsteady flow equation.

10. Belanger Equation can be written as

$$\frac{\partial H}{\partial x} = -\frac{f}{D}\frac{V^2}{2g} - S_f \tag{3.10}$$

where H is the total head,

x is the distance along the channel bed,

f is the Darcy friction coefficient,

V is the mean flow velocity,

g is the gravitational acceleration.

Equation is valid for a steady open channel flows. It is the basic equation for hydraulic jump in differential form.

3.2.2. Order and Degree of a Differential Equation

The order of a differential equation is determined by its highest derivative, whether it is an ordinary differential equation or a partial differential equation. For instance, we can see that Equation (3.9) is a first-order ordinary differential equation. A partial differential equation of the second order, the Laplace equation is written as Equation (3.7).

$$k\frac{\partial \phi}{\partial t} = \frac{\partial^2 \phi}{\partial x^2} \tag{3.11}$$

Equation (3.11) is a partial differential equation of the second order and of degree one. It is the equation of heat flow in one dimension. The following equation for transverse vibration of an elastic rod is a fourth- order linear partial differential equation.

$$\frac{\partial^2 \phi}{\partial t^2} + \alpha^2 \frac{\partial^4 \phi}{\partial x^4} = 0 \tag{3.12}$$

3.2.3. Linear and Non-Linear Differential Equations

If the dependent variable of an equation and its derivatives are first degree, that equation is a linear equation. The most general linear ordinary differential equation of order n is of the form

$$P_0\frac{d^n\varphi}{dx^n} + P_1\frac{d^{n-1}\varphi}{dx^{n-1}} + P_2\frac{d^{n-2}\varphi}{dx^{n-2}} + P_{n-1}\frac{d\varphi}{dx} + P_n\varphi + P_{n-1} = 0 \tag{3.13}$$

where P_0, P_1, ..., P_{n+1} are either constants or only functions of the independent variable x.

The simple wave equation

$$\frac{\partial A}{\partial t} + \frac{\partial Q}{\partial x} = 0 \tag{3.14}$$

or

$$\frac{\partial Q}{\partial t} + c\,\frac{\partial Q}{\partial x} = 0 \tag{3.15}$$

is linear, provided the celerity of the wave *c* is constant. The following equation is non-linear.

$$\frac{\partial u}{\partial t} + u\frac{\partial u}{\partial x} + g\frac{\partial h}{\partial x} = g(S_0 - S_f) \tag{3.16}$$

with

$$S_f = \frac{n^2u^2}{R^{4/3}} = \frac{u^2}{C^2R} \tag{3.17}$$

where n is the Manning friction coefficient, C is the Chezy friction coefficient.

Additionally, the equation is quasi-linear if the coefficients P_0, P_1, P_2, and P_{n+1} are functions of both the dependent and independent variables as well as the lower-order derivatives. The momentum equation for a one-dimensional unsteady gradually-varying flow is Equation (3.16), and it is quasi-linear. An nth-order partial differential equation is typically nonlinear when its coefficients depend on the nth-order derivatives; it is quasi-linear when their dependence is on the mth-order derivatives such that $m \le n$.

3.2.4. Classification of a Second-Order Partial Differential Equation

The conservation laws in physics and engineering can be described by a partial differential equation of the form

$$a\frac{\partial^2 \varphi}{\partial x^2} + b\frac{\partial^2 \varphi}{\partial x\,\partial y} + c\,\frac{\partial^2 \varphi}{\partial y^2} + d\,\frac{\partial \varphi}{\partial x} + e\,\frac{\partial \varphi}{\partial y} + f\varphi + g = 0 \tag{3.18}$$

in which a, b, c, d, e, f and g are constants.

Hyperbolic PDE: If $(b^2 - 4ac) > 0$, the equation is hyperbolic. The hyperbolic partial differential equations describe the propagation of a wave, which is an unsteady flow. There are two real and different roots to a hyperbolic partial differential equation. The wave equation, Equation (3.5), is a hyperbolic partial differential equation.

Parabolic PDE: If $(b^2 - 4ac) = 0$, the equation is parabolic. The diffusion or decay of a substance connected to a diffusion coefficient D is described by the parabolic PDEs.

The flow is unsteady as well. A parabolic PDE has two real and equal roots. The diffusion or heat conduction equation, Equation (3.9), is a parabolic partial differential equation.

Elliptic PDE: If $(b^2 - 4ac) < 0$, the equation is elliptic. Problems in groundwater are represented by elliptic PDEs. The flow is steady. There are two imaginary and different roots to an elliptic partial differential equation. The Laplace equation, Equation (3.7), is an elliptic partial differential equation.

3.3. Solution of Emptying Time of a Dam Reservoir

a. Derive an equation between emptying time vs water level for zero inflow case, for Göktürk dam reservoir located in Istanbul
b. Estimate a relationship between reservoir storage vs water level such as $S = ah^b$.
c. Compute the time needed to lower the water level from 10 m depth to 2 m or to empty the reservoir,
d. Plot a graph between emptying time vs water level.

Solution:

a. Derivation of an emptying time equation:

The calculation of emptying time of a dam reservoir can be derived from the conservation of mass and energy conservation equations. This relation will be derived here.

Conservation of Mass: We ignore the evaporation and seepage losses when emptying the reservoir. It has also been acknowledged that the water entering the reservoir is far smaller than the water leaving it, and may thus be neglected (See Figure 3.1). Equation (3.1) is expressed as follows if the inflow into the reservoir is zero (Inflow, I = 0).

$$-O = \frac{dS}{dt} \tag{3.19}$$

where O is the outflow,

S is the storage volume,

t is the time.

The minus sign indicates that the volume is decreasing during the emptying process.

As can be seen in Figure 3.1, the water only exits from an outlet (orifice) at the bottom of the dam, and the flow rate that outflows from the orifice at any time is written from the continuity equation as follows,

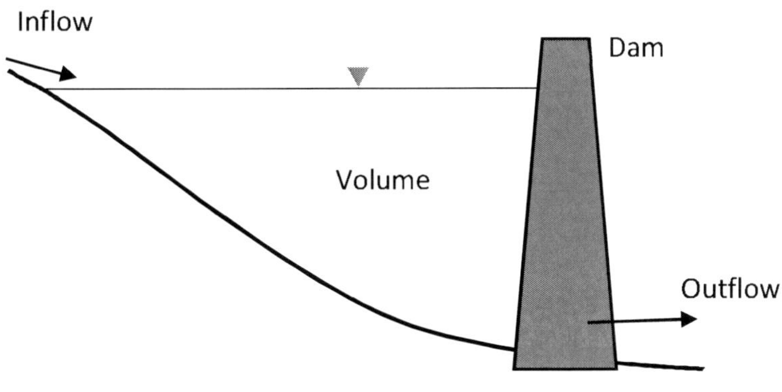

Figure 3.1. Water balance (mass conservation) in a dam reservoir.

$$O = vA \tag{3.20}$$

where O is the outflow,

A is the outlet cross sectional area,

v is the mean velocity at the orfice.

Substituting Equation (3.20) in Equation (3.19) yields,

$$\frac{dS}{dt} = -Av \tag{3.21}$$

To complete the formulation of the problem, the flow rate needs to be determined.

Conservation of energy: Let's consider kinetic and potential energies in the movement of a small body of water from the water surface in the reservoir to the axis of the orifice. From the conservation of energy principle, kinetic energy will be equal to potential energy and the following relation can be written as,

$$\frac{1}{2}mv^2 = mgh \tag{3.22}$$

where m is the mass,
v is the mean velocity in the orifice,
h is the water depth form water surface to the outlet,
g is the gravitational acceleration.
From Equation (3.22) we can write

$$v^2 = 2gh \tag{3.23}$$

or

$$v = \sqrt{2gh} \tag{3.24}$$

Using Equation (3.20) and adding a discharge coefficient C, the flow rate of an orifice can be written as:

$$Q = AC\sqrt{2gh} \tag{3.25}$$

Using Equation (3.21) and Equation (3.25), we can combine them,

$$AC\sqrt{2gh} = -\frac{dS}{dt} \tag{3.26}$$

$$dt = -\frac{dS}{AC\sqrt{2gh}} \tag{3.27}$$

If the power function of the storage-elevation curve is known as:

$$S = ah^b \tag{3.28}$$

$$dS = abh^{b-1}\,dh \tag{3.29}$$

$$dt = -abh^{b-1}\,dh/(AC\sqrt{2gh}) \tag{3.30}$$

or

$$dt = -abh^{b-1}dh/(ch^{d}) \tag{3.31}$$

The general formula is obtained as:

$$\int_0^T dt = -\frac{ab}{c}\int_{h1}^{h2}(h^{b-1-d})\,dh \tag{3.32}$$

As a conclusion the solution of emptying time of a reservoir is

$$T = \frac{ab}{c(b-d)}(h_2^{\,b-d} - h_1^{\,b-d}) \tag{3.33}$$

where $c = AC\sqrt{2g}$, $d = 0.5$ for orifice outflow, a and b are coefficients for power law function.

b. Estimate a relationship between reservoir storage vs water level such as $S = ah^b$.

The elevation and reservoir area data for Istanbul, Göktürk Dam, used in this exercise is given in Table 3.1.

Table 3.1. Elevation-area-volume curve values for Göktürk Dam

Elevation (m)	Water depth (m)	Reservoir area (m^2)	Average area (m^2)	Elevation difference (m)	Volume increase (m^3)	Reservoir volume (m^3)
44	0	0	0	0	0	0
46	2	57500	28750	2	57500	57500
48	4	85000	71250	2	142500	200000
50	6	140000	112500	2	225000	425000
52	8	195000	167500	2	335000	760000
54	10	273500	234250	2	468500	1228500
56	12	307500	290500	2	581000	1809500
58	14	390000	348750	2	697500	2507000

Solution: A power law function that is in the form of Equation (3.28) is fitted to the scatter diagram of reservoir elevation-volume data (h-S) using least square method.

The coefficient a and b are found as a = 14010, b =1.9435.
So, the elevation-volume curve can be given as:

$S = 14010 \times h^{1.9435}$

Table 3.2 and Figure 3.2 show the measured and calculated storage volumes vs elevation values.

Table 3.2. Measured and Calculated storage volumes

Elevation, h	Measured Storage volume, S	Calculated storage volume S
(m)	(m^3)	(m^3)
0	0	0
2	57500	53888
4	200000	207273
6	425000	455801
8	760000	797248
10	1228500	1230094
12	1809500	1753182
14	2507000	2365582

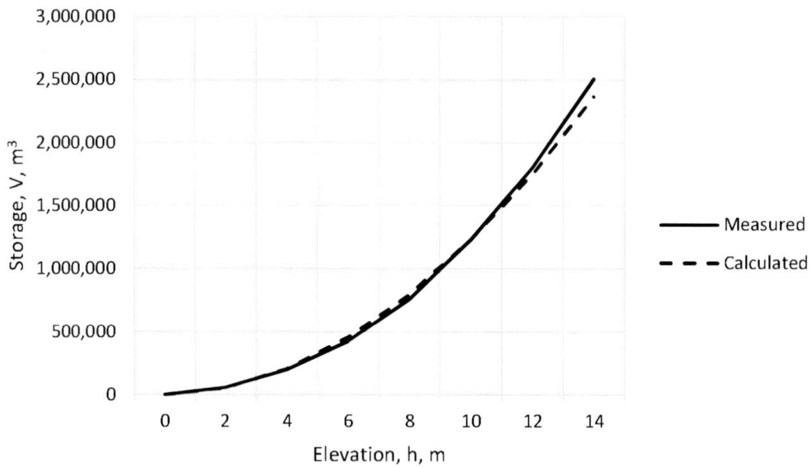

Figure 3.2. Elevation –Storage Volume curve for Göktürk Dam reservoir in İstanbul.

c. Compute the time needed to lower the water level from 10 m depth to 2 m or to empty the reservoir.

Given data:
Discharge coefficient: C = 0.63,
Orifice diameter: D = 0.60 m,
Initial depth: h_2 = 10 m,
Final depth: h_l (any level).
Find: What is the emptying time, T.
Solution: Using the given data, we find the coefficients as

$$c = AC(2g)^{0.5} = (0.6\times0.6\times3.14/4)\times0.63\times(19.62)^{0.5} = 0.789$$

$$d = 0.5$$

Therefore, the following equation represents the reservoir's emptying time:

$$T = [\ (14010\times1.9435)/(0.789)\times1.4435]^{*}\ (h_2^{1.4435} - h_1^{1.4435})$$

$$T = 23907.21(h_2^{1.4435} - h_1^{1.4435})$$

For different h_l values as 8, 6, 4, 2, and 0; emptying time is calculated using Excel spreadsheet and are shown in Table 3.3 and in Figure 3.3.

As seen form Table 3.3 and Figure 3.3, if initial depth is 10 m, and final depth is 2 m, the emptying time of the reservoir is calculated as 6.9 days.

For final water depth of zero, the emptying time is calculated 7.7 days as can be seen from Table 3.3.

Table 3.3. Calculation of reservoir emptying time using a spreadsheet

Initial depth, m	Final depth, m	Emptying Time, T		
h_2	h_l	seconds	hours	days
10	10	0	0	0
10	8	182793.6	50.8	2.1
10	6	346253.2	96.2	4
10	4	486938.3	135.3	5.6
10	2	598764.4	166.3	6.9
10	0	663787.2	184.4	7.7

d. Plot a graph between emptying time vs water level: Using the values in Table 3.3, the following graph can be plotted as shown in Figure 3.3.

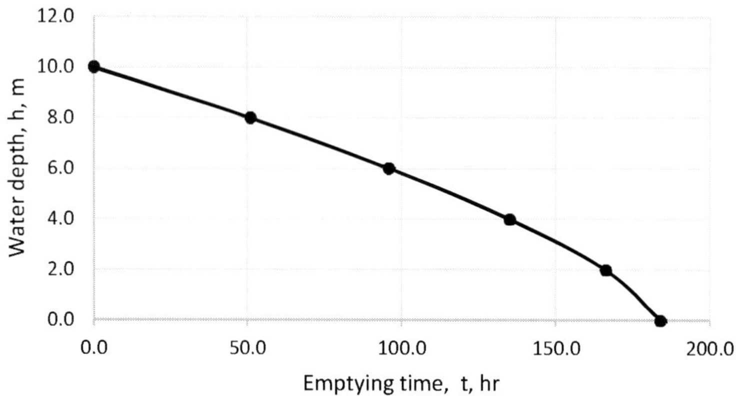

Figure 3.3. Reservoir emptying time vs water level.

3.4. Solution of an Ordinary Linear Differential Equation

Let us consider an ordinary differential equation such as shown in Equation (3.34),

$$\frac{dy}{dt} + Py = 0 \tag{3.34}$$

where y is the variable and P is either numeric constants or functions of x. Here, we can use separation of variables method to find the analytical solution for Equation (3.34). Once the equation is rewritten as each of two variables appear on the left and right hand sides of the equation, one can integrate both sides as following.

$$\int \frac{dy}{y} = \int -Pdt \tag{3.35}$$

$$\ln y = -Pt + \ln C \tag{3.36}$$

$$y = Ce^{-Pt} \tag{3.37}$$

where, C is a constant that can be obtained from a given initial condition.

At an initial moment, i.e., t = 0, suppose the concentration of y is equal to y_0 (initial value of the variable). Any fixed value or distribution with respect to x, i.e., $y_0(x)$, can be used as the initial concentration. This initial condition allows us to calculate the following expression.

$$y_0(x) = C \tag{3.38}$$

Therefore, the analytical solution for Equation (3.34) can be obtained as

$$y(x, t) = y_0(x)e^{-Pt} \tag{3.39}$$

Here, we can consider two different cases such as constant C_0 and variable C_0 as adapted from Ali (2014).

Table 3.4. Variation of concentration with time: Analytical solution for constant C_0

Time		Relative values
t	Pt	y/y_0
0	0	1
0.1/P	0.1	0.9048
0.2/P	0.2	0.8187
0.3/P	0.3	0.7408
0.4/P	0.4	0.6703
0.5/P	0.5	0.6065
1/P	1	0.3679
2/P	2	0.1353

Table 3.5. Variation of concentration with time: Analytical solution for y_0 that follows a Gaussian distribution (in which σ is standard deviation and taken as $\sigma = 0.25$)

x	t = 0.0	t = 1.0
	y/y_0	y/y_0
-1	0.0005	0.000184
-0.75	0.0018	0.000662
-0.5	0.2165	0.07965
-0.25	0.9704	0.35701
0	1.6	0.58864
0.25	0.9704	0.35701
0.5	0.2165	0.07965
0.75	0.0018	0.000662
1	0.0005	0.000184

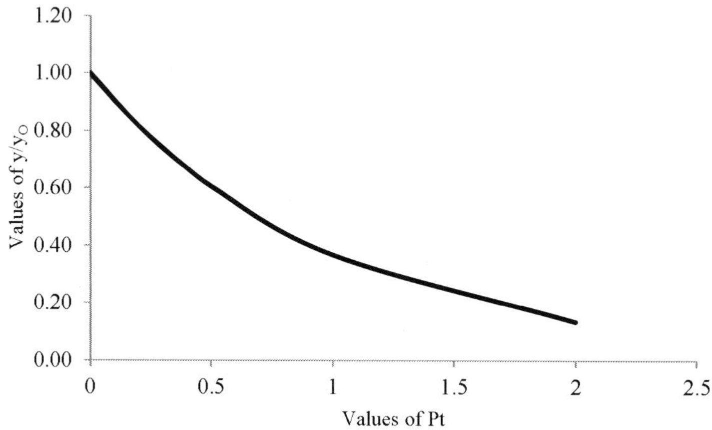

Figure 3.4. Analytical solution graph for linear differential equation.

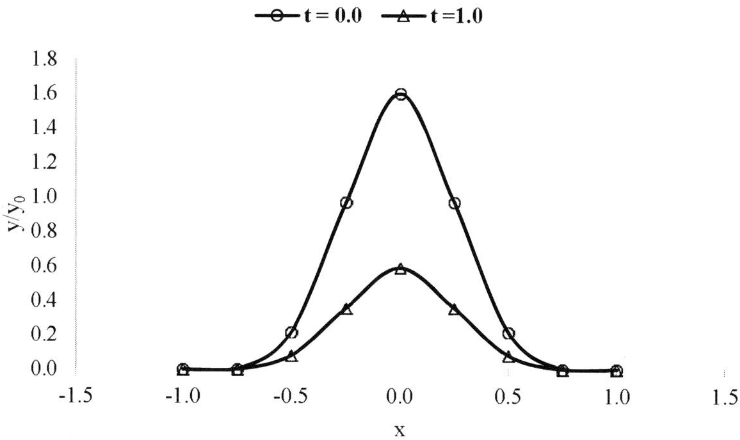

Figure 3.5. Decay of a Gaussian distribution with time for $\sigma = 0.25$.

Case 1: $y_0(x) = C$ Constant Value of y_0:

If initial concentration is taken as a constant value of y_0, the analytical solution can be given as shown in Table 3.4 and Figure 3.4. The results show that there is a decay of this y_0 with time.

Case 2: $y_0(x)$ = A Gaussian Distribution:

Now let us take the initial concentration $y_0(x)$ is a Gaussian function such as:

$$y_0(x) = y(x, 0) = \frac{1}{\sigma\sqrt{2\pi}} e^{-\frac{1}{2}\left(\frac{x-\mu}{\sigma}\right)^2} \tag{3.40}$$

where, μ and σ are the mean and standard deviation of the distribution, respectively. In this example μ is taken as 0. The variation of concentration with time for this initial condition are shown in Table 3.5 and in Figure 3.5.

3.5. Example for a Simple Lake Purification Model

The purification model will be applied to the Eğridir Lake, in Turkey. The volume of this lake is $V = 4.0 * 10^9$ m^3 and yearly average outflow rate $r = 1.4 \times 10^6$ m^3/day.

Find how much time is needed before the level of pollution is reduced to 10% of its initial value.

Solution: Consider that a population of particles contaminate the lake which has a constant volume, V. We assume that the pollutant concentration is uniform throughout the lake. Let us take c(t) is the concentration of the pollutant at time t. Since the lake volume is constant, the outflow, r, from the lake can be taken as same as difference between rainfall and evaporation.

The relevant variables and parameters are as c(t), V and r. By using the mass balance equation (volume change = Inflow volume – Outflow volume), the state equation of pollution concentration can be written as:

Rate of change of pollution = Pollution inflow - Pollution outflow or mathematically, (c(t) is written as c for simplicity) (Ali, 2014; Chanson, 2004),

$$\frac{d}{dt}(cV) = 0 - rc \tag{3.41}$$

where c is the concentration of the pollutant,

V is the lake volume,

r is the outflow rate from the lake,

t is the time.

Since V is constant, Equation (3.41) becomes,

$$d(c)/c = -(r/V)dt \tag{3.42}$$

Now, by separation of variables and direct integration it can be found as:

$$c(t)/c(0) = e^{-rt/V} \quad (3.43)$$

To give a answer to the question asked above, solve for c(t)/c(0) = 0.10,

$$t = -\ln(0.10)\times(V/r) \quad (3.44)$$

This result can be used for Eğridir Lake, in Turkey. This lake has $V = 4.0 \times 10^9$ m^3 and yearly average outflow is $r = 1.4\times10^6$ m^3/day. Substituting these values, it is calculated that t = time ≈ 18 years.

The rate of the decay is actually depended on the negative exponent, -r/V < 0. Different decay models are presented in Table 3.6 and Figure 3.6.

In reality, it is obvious that more complex purification process occurs depending upon the varying r and c(t). For instance, r could change through out the year and non-uniform concentration could be defined as c (x, y, t). To take all the realistic assumptions into consideration, a numerical model can be established that assess both temporal and spatial variations.

Also, for a numerical model boundary conditions should be defined, considering the physical location of the lake. This model is the simplest version of the purification model.

Table 3.6. Calculation of c/c(0) versus time (t) for different rate of flows

Time, t,	Rate of outflows from the lake, r			
Year	1.4×10^6 m^3/day	1.2×10^6 m^3/day	1.0×10^6 m^3/day	0.8×10^6 m^3/day
	c/c(0)			
0	1	1	1	1
1	0.88	0.9	0.91	0.93
2	0.77	0.8	0.83	0.86
3	0.68	0.72	0.76	0.8
4	0.6	0.65	0.69	0.75
5	0.53	0.58	0.63	0.69
6	0.46	0.52	0.58	0.65
7	0.41	0.46	0.53	0.6
8	0.36	0.42	0.48	0.56
9	0.32	0.37	0.44	0.52
10	0.28	0.33	0.4	0.48
12	0.22	0.27	0.33	0.42
14	0.17	0.22	0.28	0.36
16	0.13	0.17	0.23	0.31
18	0.1	0.14	0.19	0.27
20	0.08	0.11	0.16	0.23

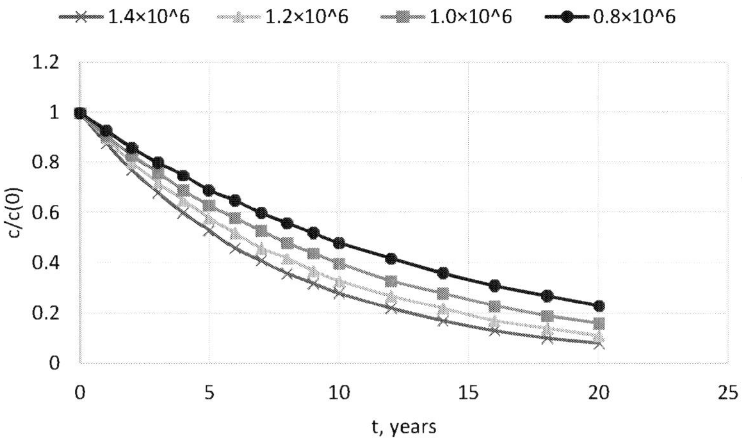

Figure 3.6. Reduction of concentration of the pollutant versus time for different rate of outflows from Eğridir Lake.

Chapter 4

Applications of Numerical Analysis

4.1. Introduction

Numerical analysis methods are a way to solve civil as well as other engineering problems, physical or mathematical problems using computers. These numerical techniques have become the field that engineers and scientists frequently use to solve their unique problems (Hoffman, 2001; McDonough, 2008). The biggest advantage of these methods is that even the problems that do not have an analytical solution or which take a lot of time to solve analytically can be obtained with these methods. Analytical methods often produce solutions in the form of mathematical functions. These general solutions are then converted into numerical results for some specific data. In numerical solutions, the results are always numerical. Another important difference of numerical solutions is that they produce approximate solutions. However, these approximate solutions can be obtained to the extent desired. The process steps increase as the accuracy is increased, but today's computers, which are very fast, can come to a conclusion in a short time in the face of increasing process steps. Sometimes even in analytical solutions, the same level of errors can be found in calculations made to obtain the same values.

In fact, the history of numerical algorithms is as old as human civilization. In ancient Egypt, on Rhind Papirus, in 1650 BC, a method of finding a root for solving a simple equation was described. Al-Khwarizmi proposed the solution of quadratic equations using the perfect square method in the 820s. He is known as the founder of algebra. Also, his name gave rise to term 'algorithm.' The theory of numerical methods was introduced in the 1930s, but this calculation was not easy to do manually. However, with the development of computers, these methods became easier to apply.

In the solution of mathematical models, with the increasing importance of the use of computers to perform numerical operations, scientific computing or computational science took shape in the 1980s and 1990s. This field was viewed as numerical analysis from a computer science perspective. Thus, real-life mathematical modeling was better understood by using numerical analysis, computer graphics, and mathematical computations (Popescu, 2014).

The main topics of numerical analysis and the topics that are tried to be summarized below are included in this section: Finding the roots of equations, solutions of linear and nonlinear equation sets, finding internal values (interpolation) and external values (extrapolation) using a series of data, fitting a curve suitable for a number of known values and measurement points obtained in experimental studies, integration of functions or points from a series of experimental measurements.

4.2. Newton-Raphson Method

There are some difficulties in finding a root of a complicated algebraic function. There are ways to numerically find the roots of complicated functions using some basic concepts of mathematics. The Newton-Raphson method is commonly used for this purpose. In this sequential process, a series of operations are followed in finding an approximate root, taking into account the function, its derivative and its initial value.

As will be remembered from algebra, the root of a function is the value that makes the function zero. As seen in Figure 4.1, the derivative of a function can be written as follows.

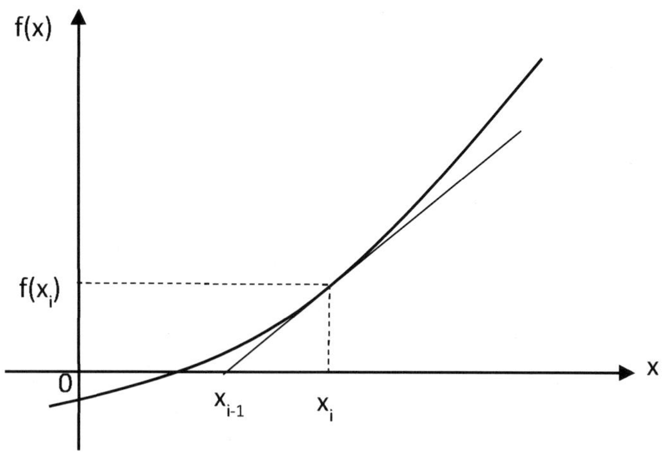

Figure 4.1. Schematic representation of Newton-Raphson Method.

$$\frac{df}{dx} = f'(x_i) = \frac{f(x_i) - 0}{x_i - x_{i+1}} \tag{4.1}$$

Rearranging this equation, one can obtain

$$x_{i+1} = x_i - \frac{f(x_i)}{f'(x_i)} \tag{4.2}$$

As previously mentioned, the Newton -Raphson method is an iterative process used to find the root of a function. The specific root determined by the process depends on the initial state and the randomly chosen x value.

$$x_{n+1} = x_n - \frac{f(x_n)}{f'(x_n)} \tag{4.3}$$

here, x_n is the current value of x,

$f(x_n)$ is the value of the function at x_n,

$f'(x_n)$ is the derivative at x_n,

x_{n+1} shows the next value tried to find for x.

In fact, $f'(x)$ shows derivative as $f(x)/\Delta x$. Therefore, $f(x)/\ f'(x)$ represents a value of Δx.

$$x_{n+1} - x_n = \frac{f(x)}{f'(x)} = \frac{f(x)}{f(x)/\Delta x} = \Delta x \tag{4.4}$$

4.3. Least Squares Method

In least squares method, so-called numerical approximation and interpolation or regression analysis, it is assumed that a series of x_i values and the corresponding $y_i = f_i(i=1,.....\ n)$ are given. The interpolation is defined on approximate computation of the f value for $x \neq x_i$. Such series of points (f_i, x_i) may arise from laboratory experiments or may be given in figure or table describing the use of an instrument or a monograph (Figure 4.2).

When incontrollable errors are included in the f_i values (on with laboratory experiments), then the curve passing from all points on the f-x plane may not be smooth and the cloud of point is usually approximated by a curve passing among the point in a weighted way.

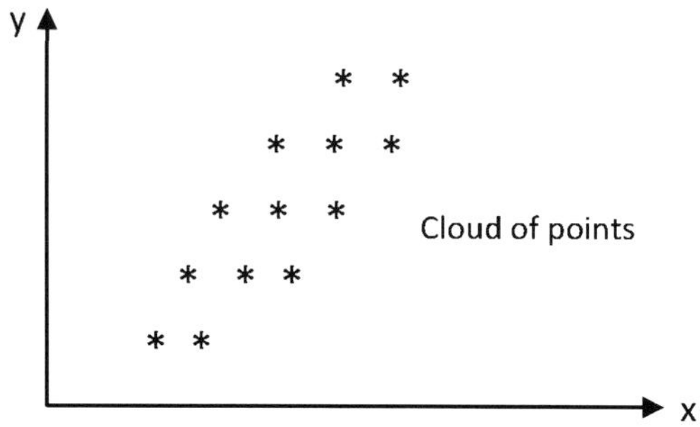

Figure 4.2. Experimental Cloud of Points Graph or scatter diagram.

This is usually done by the least squares method, achieving a minimization of the variance between the smooth curve and the experimental points. This analysis is useful in finding the "best fit" of a set of experimental observations to a mathematical relationship. If the best straight line is sought, then it is linear regression analysis (Figure 4.3).

The independent variable is x and this is assumed to be error free. The dependent variable is y and this is the result of experimental observation and thus contains errors.

The general straight line is:

$$y = mx + c \tag{4.5}$$

where m is the slope of the line and c is the intercept of the y axis (Figure 4.4).

Suppose we have n values of x_i and corresponding observation y_i. We use the principle of "Least-square" to find the best straight line to represent the relationship between y and x.

The principles state that the sum of the squares of the "errors" or "deviations" should be a minimum, i.e., $\sum(y_i - y)^2$ should be a minimum. Now,

$$R = \sum_{i=1}^{n} (y - y_i)^2 = \sum_{i=1}^{n} (mx_i + c - y_i)^2 = \min \tag{4.6}$$

The errors may be negative or positive so the squares of errors are always positive. The minimization in Eq. 4.6 is achieved by making the derivatives zero,

$$\frac{\partial R}{\partial m} = 0 \quad (4.7)$$

$$\frac{\partial R}{\partial c} = 0 \quad (4.8)$$

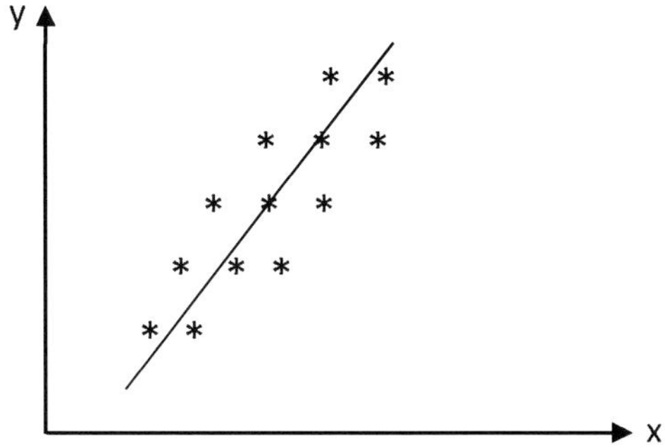

Figure 4.3. Linear regression analysis graph.

Differentiating the right-hand side of equation partially with respect to *m* and *c* we obtained:

$$R = \sum_{i=1}^{n} (mx_i + c - y_i)^2 \quad (4.9)$$

$$\frac{\partial R}{\partial m} = \sum 2(mx_i + c - y_i)(x_i) = \sum_{i=1}^{n}\left(mx_i^2 + cx_i - y_i x_i\right) = 0 \quad (4.10)$$

$$c\sum_{i=1}^{n} x_i + m\sum_{i=1}^{n} x_i^2 = \sum_{i=1}^{n} x_i y_i \quad (4.11)$$

$$\frac{\partial R}{\partial c} = 2\sum_{i=1}^{n} (mx_i + c - y_i) = 0 \quad (4.12)$$

$$nc + m\sum_{i=1}^{n} x_i = \sum_{i=1}^{n} y_i \quad (4.13)$$

Solving Equations (4.11) and (4.13) yields the values of *m* and *c* where the line $y = mx + c$ is the line of regression of *y* vs. *x* (Figure 4.4).

$$nc\sum x_i + nm\sum x_i^2 = n\sum x_i y_i \quad (4.14)$$

$$-nc\sum x_i - m\sum x_i \sum x_i = -\sum x_i \sum y_i \quad (4.15)$$

If you sum up Equations (4.14) and (4.15) side by side, you will obtain:

$$m(n\sum x_i{}^2 - (\sum x_i)(\sum x_i)) = n\sum x_i y_i - \sum x_i \sum y_i \tag{4.16}$$

$$m = \frac{n\sum_{i=1}^{n}(x_i y_i) - \sum_{i=1}^{n} x_i \sum_{i=1}^{n} y_i}{n\sum_{i=1}^{n} x_i^2 - (\sum_{i=1}^{n} x_i)^2} \tag{4.17}$$

and

$$c = \frac{\sum_{i=1}^{n} x_i{}^2 \sum_{i=1}^{n} y_i - \sum_{i=1}^{n} x_i \sum_{i=1}^{n}(x_i y_i)}{n\sum_{i=1}^{n} x_i{}^2 - (\sum_{i=1}^{n} x_i)^2} \tag{4.18}$$

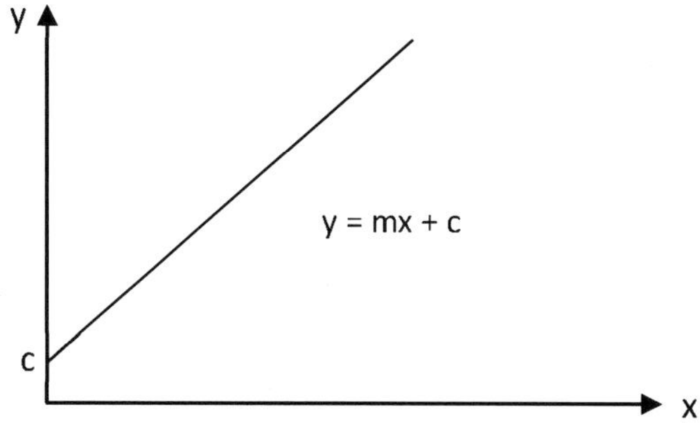

Figure 4.4. Linear equation graph.

4.4. Polynomial Curve Fitting

There are many instances in civil engineering where it is useful to fit a mathematical relationship, in the form of a polynomial, to experimental data. The general form of the polynomial is taken to be:

$$y = a_0 + a_1x_i + a_2x_i{}^2 + \ldots + a_nx_i^n \tag{4.19}$$

The largest power of x donates the order of the polynomial. When n = 1 the first degree polynomial) is a straight line and if n = 2, we will have the quadratic $y = a_1 + a_1x + a_2x^2$ relationship between *y* and x. If it is assumed that the intervals x, at which observed values of *y* are available, are "exact" then any errors are confined to the observations y. If any observed values y_i is known, the corresponding "residual," Δ_i, is expressed as below:

$$\Delta_i = y_i - (a_0 + a_1x_i + a_2x_i{}^2 + \ldots + a_nx_i^n) \tag{4.20}$$

Fitting of a polynomial involves choosing values for the coefficients a_0 > a_1....a_n which together minimize the sum of the squares of all the residuals as given by equations such an Eq. (4.21). If there are p observed values of y and if E is the sum of the squares of residuals that must be minimized.

$$E = \sum_{i=1}^{p}\{y_i - (a_0 + a_1x_i + a_2x_i^2 + \ldots + a_xx_i^n)\}^2 = \sum_{i=1}^{p}\Delta_i^2 \qquad (4.21)$$

The proceeding is to differentiate Eq. (4.21) with respect to the coefficients a_0, a_1, ..., a_n in turn, and equate to zero. The resulting equations are called the "normal" equations and their solution gives the "best" values of the coefficient in the polynomial of the chosen order *n*. Suppose *n* = 2, then, the sum of the squares of residuals can be minimized as below.

$$\Delta_i = y_i - (a_0 + a_1x_i + a_2x_i^2) \qquad (4.22)$$

$$E = \sum_{i=1}^{p}\Delta_i^2 = \sum_{i=1}^{p}(y_i - (a_0 + a_1x_i + a_2x_i^2)^2 \qquad (4.23)$$

$$\frac{\partial E}{\partial a_0} = -2y_i + 2a_0 + 2a_1x_i + 2a_2x_i^2 = 0 \qquad (4.24)$$

$$\frac{\partial E}{\partial a_1} = -2x_iy_i + 2a_0x_i + 2a_1x_i^2 + 2a_2x_i^3 = 0 \qquad (4.25)$$

$$\frac{\partial E}{\partial a_2} = -2x_i^2y_i + 2a_0x_i^2 + 2a_1x_i^3 + 2a_2x_i^4 = 0 \qquad (4.26)$$

These lead to the equations

$$pa_0 + a_1x_i + a_2\sum x_i^2 = \sum y_i \qquad (4.27)$$

$$a_0\sum x_i + a_1\sum x_i^2 + a_2\sum x_i^3 = \sum x_iy_i \qquad (4.28)$$

$$a_0\sum x_i^2 + a_1\sum x_i^3 + a_2\sum x_i^4 = \sum x_i^2y_i \qquad (4.29)$$

In each case the summation is taken over the p observation. These are the normal equations corresponding to the second order polynomial. In matrix form, normal equations can be represented as

$$\begin{bmatrix} p & \sum x_i & \sum x_i^2 \\ \sum x_i & \sum x_i^2 & \sum x_i^3 \\ \sum x_i^2 & \sum x_i^3 & \sum x_i^4 \end{bmatrix} \begin{bmatrix} a_0 \\ a_1 \\ a_2 \end{bmatrix} = \begin{bmatrix} \sum y_i \\ \sum x_i y_i \\ \sum x_i^2 y_i \end{bmatrix} \tag{4.30}$$

We can also obtain the matrix for the general case of a polynomial of order *n* as below.

$$\begin{bmatrix} p & \sum x_1 \cdots & \sum x_i^n \\ \sum x_i & \sum x_i^2 \cdots & \sum x_i^{n+1} \\ \vdots & \vdots & \vdots \\ \sum x_i^n & \sum x_i^{n+1} \ldots & \sum x_i^{n+n} \end{bmatrix} \begin{bmatrix} a_0 \\ a_1 \\ \vdots \\ a_2 \end{bmatrix} = \begin{bmatrix} \sum y_i \\ \sum x_i y_i \\ \vdots \\ \sum x_i^n y_i \end{bmatrix} \tag{4.31}$$

4.5. Numerical Integration

In mathematics, definite integral is used to find the area under a curve. In engineering applications, there are some physical interpretations or explanations of finding the area under a curve. For example, the area under a flood hydrograph represents the total volume of the water passed the cross section of a river during the flood event. When we approximate integral using numerical methods, it's called *numerical integration.* Several computational schemes exist for numerical integration and can be derived from Newton's formula, such as Trapezoidal Rule and Simpson's Rule are explained in this section.

4.5.1. Numerical Integration Using Trapezoidal Rule

Assume the integral given below:

$$A = \int_{x_0}^{x_1 = x_0 + h} f(x)\,dx \tag{4.32}$$

Trapezoidal Rule works by approximating the region under the graph of the function $f(x)$ (see Figure 4.5) as one or more trapezoids and calculating their area.

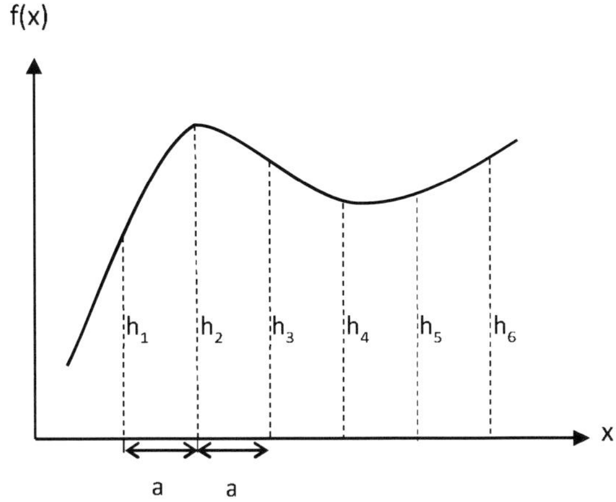

Figure 4.5. Scheme for trapezoidal rule.

Considering Figure 4.5, the area between the curve and the *x* axis is obtained approximately by adding the areas of the trapezoidal formed by drawing chords between the ordinate. Thus, the function (curve) is replaced by a sequence of straight lines between each successive pair of ordinates. The general formula is given in Equation 4.33.

$$A = \frac{a}{2}(h_1 + h_2) + \frac{a}{2}(h_2 + h_2) + \cdots + \frac{a}{2}(h_{n-1} + h_n) \tag{4.33}$$

4.5.2. Numerical Integration by Simpson's Rule

A more accurate evaluation of area may generally be obtained using Simpson's rule. In this rule, the function is replaced by a parabolic segment between successive groups of three ordinates as shown in Figure 4.6. The upper part of the closed area in the figure is therefore equals to $\frac{4}{3}ay_c$. Thus, the total area between the curve and the *x* axis is calculated using Equation 4.34.

$$A = a\left\{\frac{1}{2}(h_1 + h_2 - y_c) + \frac{1}{2}(h_2 - y_c + h_3) + \frac{4}{3}ay_c\right\} \tag{4.34}$$

Substituting for

$$y_c = h_2 - \frac{1}{2}(h_1 + h_3) \text{ and } A = \frac{a}{3(h_1 + 4h_2 + h_3)} \tag{4.35}$$

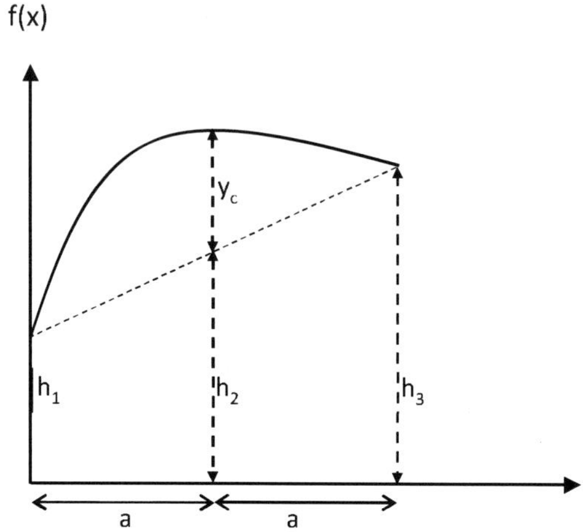

Figure 4.6. Area of parabolic segment in Simpson's rule approximation.

In the general case with n coordinates (n-1 intervals), the total area is

$$A = \frac{a}{3}\{(h_1 + 4h_2 + h_3) + (h_3 + 4h_4 + h_5) + \cdots\} \quad (4.36)$$

$$A = \frac{a}{3}[h_1 + 4(h_2 + h_4 + h_6 + \cdots) + 2(h_3 + h_5 + h_7 + \cdots) + h_n] \quad (4.37)$$

4.6. Applications by Numerical Analysis

4.6.1. Calculation of Flow Velocity at Downstream of a Spillway Using Newton-Raphson Method

As shown in Figure 4.7, the spillway has a span width of 18 m and the maximum depth of overflow of 11.6 m. The spillway includes four radial gates and its design discharge is 3090 m^3/s. Maximum reservoir water level is 42 m, and spillway crest level is 31.5 m. Surface level of energy dissipater is 29 m.

a. Find velocity and the water head over spillway assuming the discharge coefficient of the spillway as C = 2.10.
b. Find the flow velocity downstream of the spillway.

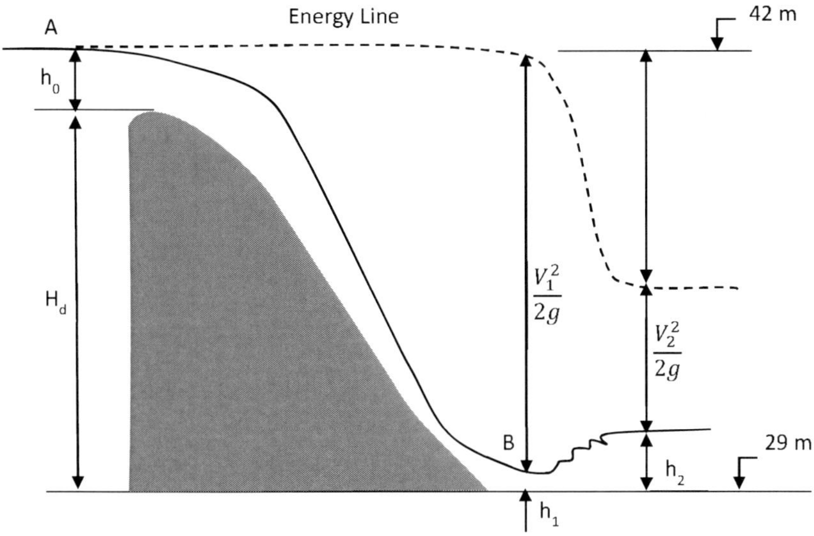

Figure 4.7. Flow over spillway.

Solution: a. The flow discharge over each span is determined by dividing the total discharge into number of gates. Therefore,

$$Q_i = \frac{3090}{4} = 772.5\ m^3/s$$

Now, empirical discharge equation can be used to obtain the head of water over spillway (h_0):

$$Q_i = Cbh_0^{1.5} = 772.5 = 2.1 \times 18 \times h_0^{1.5} \rightarrow h_0 = 7.47\ m$$

From the continuity equation the velocity over spillway is,

$$v_0 = \frac{Q_i}{bh_0} = \frac{772.5}{18 \times 7.47} = 5.75\ m/s$$

b. Of the main equations used in describing open channel flow is Bernoulli Equation. The integration of the terms obtained from the application of Newton's second law to open channel flow gives this equation. Bernoulli's equation, valid at any arbitrary point along a streamline can be written as,

$$H = \frac{v_0^2}{2g} + \frac{P_0}{\gamma} + z_0 = \frac{v_1^2}{2g} + \frac{P_1}{\gamma} + z_1$$

where: v is the flow velocity at a point on a streamline,
g is the gravitational acceleration,
z is the elevation of the point above a reference datum,
P is the relative pressure at the chosen point,
γ is the specific weight of the water,
H is the total energy of flow.

For any open channel flow ($p_A = P_0 = 0.0$ and $p_B = P_0 = 0.0$), and the velocity upstream of spillway is negligible ($v_A = 0.0$; $z_A = h_0$). Rearranging the above equation yields:

$$H = \frac{V_1^2}{2g} + h_1 = \frac{V_1^2}{2g} + \frac{Q}{V_1 b_1}$$

where: H is total flow energy,
Q is discharge over spillway,
v_1 is flow velocity downstream of the spillway,
h_1 is flow depth downstream of the spillway,
b_1 is span with of the spillway,
g is gravitational acceleration.

Considering the elevations given in the figure, the total flow energy can be obtained as following:

$$H = 42 - 29 = 13 \text{ m}$$

By substituting the values of H, b and Q_i in Equation (4.6)

$$13 = \frac{{v_1}^2}{19.62} + \frac{722.5}{18 v_1} \rightarrow v_1^3 - 255 \times v_1 + 842 = 0$$

Now, we can find the roots of this equation using Newton-Raphson method:

$$f(x) = x^3 - 255x + 842 = 0 \rightarrow f'(x) = 3x^2 - 255$$

Graphical representation of f(x) is presented in Figure 4.8.

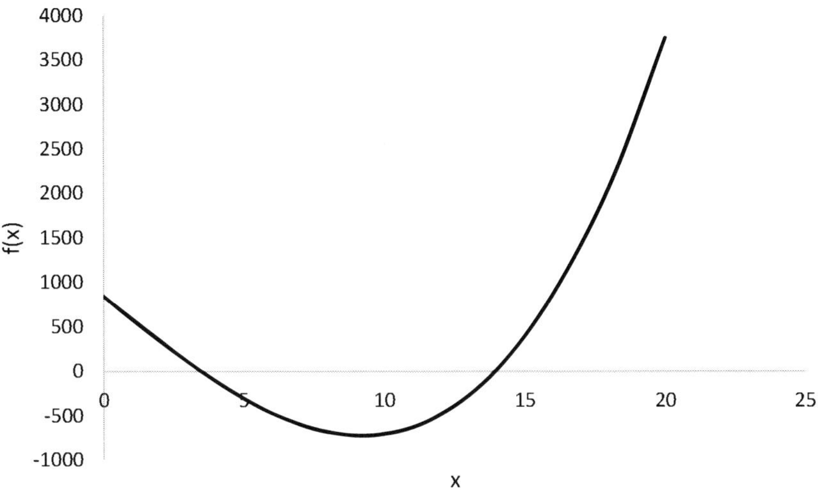

Figure 4.8. Newton-Raphson Method: for initial value as x = 5; final value of x = 3.465.

As seen from Figure 4.1, the function has two real roots. One of them is near to 5, and the other is near to 15. Firstly, we estimate the initial value as 5. We can calculate the root using Newton-Raphson method as Table 4.1.

Table 4.1. Calculation the root of function for initial value of 5, using Excel

n	x_n	$f(x_n)$	$f'(x_n)$	$f(x_n)/f'(x_n)$	x_{n+1}	Δx
0	5	-308	-180	1.711111	3.288889	1.711111
1	3.288889	38.90855	-222.55	-0.17483	3.46372	-0.17483
2	3.46372	0.306927	-219.008	-0.0014	3.465121	-0.0014
3	3.465121	2.04E-05	-218.979	-9.3E-08	3.465121	-9.3E-08

As seen from Table 4.1 one of the roots of the function is 3.465121. The flow velocity downstream of the spillway v_1 should be greater than $v_0 = 5.75$ m/s. If we consider the free fall water, velocity at the bottom should be theoretically approximate as $v_1 = (2gh)^{0.5} = (19.62\times13)^{0.5} = 16$ m/s. Thus, for the first iteration, v_1 is assumed to be 15 m/s. The other iterations are tabulated in Table 4.2.

Table 4.2 indicates that for $x_n = 13.952$, the Δx will be zero. This is the second root of the function. From the physical point of view, this is the real solution. Thus, the flow velocity downstream of the spillway is $v_1 = 13.952$ m/s.

Table 4.2. Calculation table using Newton-Raphson method

n	x_n	$f(x_n)$	$f'(x_n)$	$f(x_n)/f'(x_n)$	x_{n+1}	Δx
0	15.000	392.000	420.000	0.933	14.067	0.933
1	14.067	38.387	338.613	0.113	13.953	0.113
2	13.953	0.541	329.084	0.002	13.952	0.002
3	13.952	0.000	328.946	0.000	13.952	0.000

4.6.2. Finding a Relationship between Temperature and Liquid Volume Using Least Squares Method

Table 4.3 shows the volume of a liquid at different temperatures measured in a laboratory. Draw the regression line representing volume as the function of temperature. What would be the volume at T = 32°C (extrapolation) and T = 24.5°C (interpolation)?

Table 4.3. Volume of a liquid at different temperature

Temp. (°C)	Volume (L)	Temp. (°C)	Volume (L)
17	1.0022	22	1.0033
18	1.0024	23	1.0035
19	1.0026	24	1.0037
20	1.0028	25	1.0040
21	1.0030	26	1.0043

Solution: The outcomes of 10 experiments are listed in Table 4.3, where L stands for liter. Therefore, the number of data is n = 10. Temperature variable is denoted by x and the volume by y. The terms $\sum x_i$, $\sum y_i$, $\sum x_i{}^2$, $\sum x_i y_i$ are calculated as 215, 10.0318, 4705, and 215.703, respectively. Finally, the slope and the intercept of the regression line can be calculated using Equation (4.17) and Equation (4.18) as:

$$m = \frac{n\sum x_i y_i - \sum x_i \sum y_i}{n\sum x_i^2 - (\sum x_i)^2} = 0.00002303$$

$$c = \frac{\sum x_i{}^2 . \sum y_i - \sum x_i . \sum x_i y_i}{n\sum x_i{}^2 - (\sum x_i)^2} = 0.9982285$$

Thus, the linear relationship between volume and temperature is expressed as below:

$$V = 0.0002303\,T + 0.9982285$$

Figure 4.9 illustrates the regression line derived using Microsoft Excel.

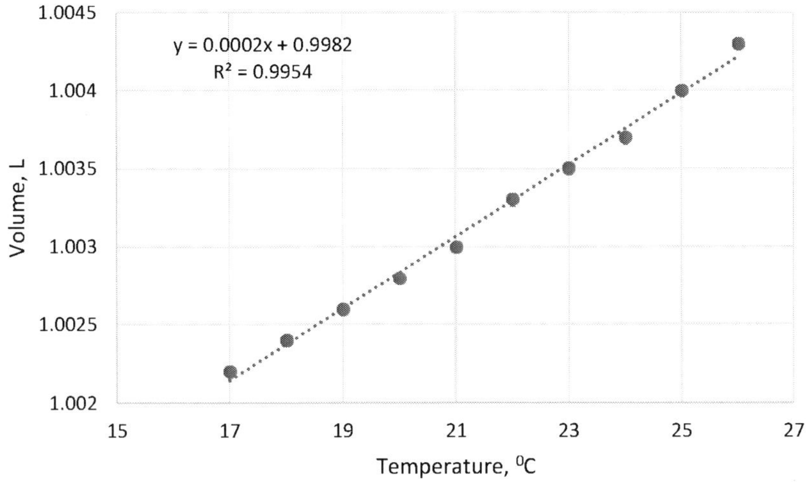

Figure 4.9. Regression line between the liquid volume and temperature.

The resultant equation can be used to find the volume via inter- or extrapolation strategies. As asked in the question. The fluid volume at 32°C and 24.5°C equals to 1.0055981 L and 1.0041012 L, respectively.

$$\text{For } x = 32\,°\text{C}, y = 0.0002303x + 0.9982285 = 1.0055981\text{ L}$$

$$\text{For } x = 25.5°\text{C}, y = 0.0002303x + 0.9982285 = 1.0041012\text{ L}$$

4.6.3. Application of Polynomial Fitting

A set of dependent *y* and independent variable *x* is given in Table 4.4. Fit a second-degree polynomial to the data.

Table 4.4. A set of dependent and independent variables

Dependent variable, x	0	1	2	3	4	5
Independent variable, y	5	15	26	27	30	35

Solution: Use the following formula as

$$y = a_0 + a_1x + a_2x^2$$

We can obtain as

$\sum x_i = 15$; $\sum x_i^2 = 55$; $\sum x_i^3 = 225$; $\sum x_i^4 = 979$; $\sum y_i = 138$; $\sum x_i^2 y_i = 1717$

Therefore, the set of equations and their matrix form are as below for p = 6, n = 2;

$$6a_0 + 15a_1 + 55a_2 = 138$$

$$15a_0 + 55a_1 + 225a_2 = 443$$

$$55a_0 + 225a_1 + 979a_2 = 1717$$

$$\begin{bmatrix} 6 & 15 & 55 \\ 15 & 55 & 225 \\ 55 & 225 & 979 \end{bmatrix} \begin{bmatrix} a_0 \\ a_1 \\ a_2 \end{bmatrix} = \begin{bmatrix} 138 \\ 443 \\ 1717 \end{bmatrix}$$

That results in: a_0= 5.61, a_1= 10.7, and a_2= -1.02. Hence, the polynomial function is

$$y = 5.61 + 10.7x - 1.02x^2$$

Figure 4.10 illustrates the 2nd order polynomial derived using Microsoft Excel.

Table 4.5 provides the corresponding y for each x using the attained polynomial curve.

Table 4.5. Experimental data and polynomial curve fitting

x	y	y *	x	y	y*
0	5	5.61	3	27	28.53
1	15	15.29	4	30	32.09
2	26	22.93	5	35	33.61

* Estimated.

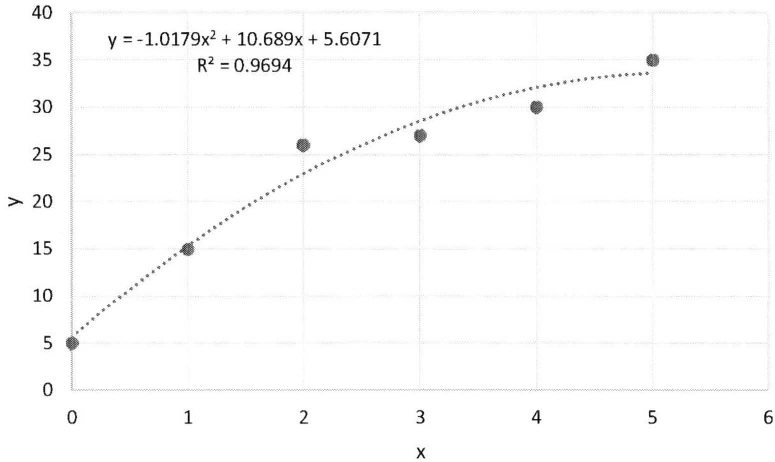

Figure 4.10. Plot of data for experimental and polynomial fitting example.

4.6.4. Determining a Relationship between Water Depth and Reservoir Volume Using Curve Fitting

Fit a power equation to the following data set presented in Table 4.6 and in Figure 4.6.

Table 4.6. Measured values of a reservoir water level and surface area

Water depth, h, m	Area, A, m^2	Reservoir Volume, V, $10^6\ m^3$
0	0	0
5	125	2.25
10	1000	8.05
15	3375	17.75
20	8000	30.5
25	15625	46
30	27000	65.7
35	42875	89.5

Solution: We assume there is a relation between water depth, h, and reservoir volume, V, as

$$V = Ah^B$$

In this equation, if the logarithm of two sides is taken, the equation becomes as,

$$\log V = \log A + B \log h$$

This equation represents a straight line equation as:

$$y = c + mx$$

The coefficients of this line, c and m, will be obtained from the following formulas using Equation (4.15) and Equation (4.16),

$$m = \frac{n\sum x_i y_i - \sum x_i \sum y_i}{n\sum x_i^2 - (\sum x_i)^2}$$

$$c = \frac{\sum x_i^2 \sum y_i - \sum x_i \sum x_i y_i}{n\sum x_i^2 - (\sum x_i)^2}$$

Calculation prepared by Excel and calculated values are presented in Table 4.7.

Table 4.7. Calculated values for curve fitting

Water depth, h, m	Reservoir Volume, V, 10^6 m^3	x = logh	y = logV	x*y	x*x
0	0			0	0
5	2.25	0.69897	0.352183	0.246165	0.488559
10	8.05	1	0.905796	0.905796	1
15	17.75	1.176091	1.249198	1.469171	1.383191
20	30.5	1.30103	1.4843	1.931119	1.692679
25	46	1.39794	1.662758	2.324436	1.954236
30	65.7	1.477121	1.817565	2.684764	2.181887
35	89.5	1.544068	1.951823	3.013748	2.384146
SUM	259.75	8.595221	7.4718	12.5752	11.0847

From the above table values, we can obtain,

m = 2.458134,

c = -1.70705,

A = 0.01963.

Therefore, the equation of the straight line is written as

$$y = mx + c$$

$$y = 2.458134x - 1.70705$$

Table 4.8. Measured and calculated reservoir volumes

Water level, h, m	Reservoir Volume, V, 10^6 m^3	
	Measured	Computed
0	0	0
5	2.25	1.025846
10	8.05	5.637081
15	17.75	15.2725
20	30.5	30.97607
25	46	53.6098
30	65.7	83.92325
35	89.5	122.5876

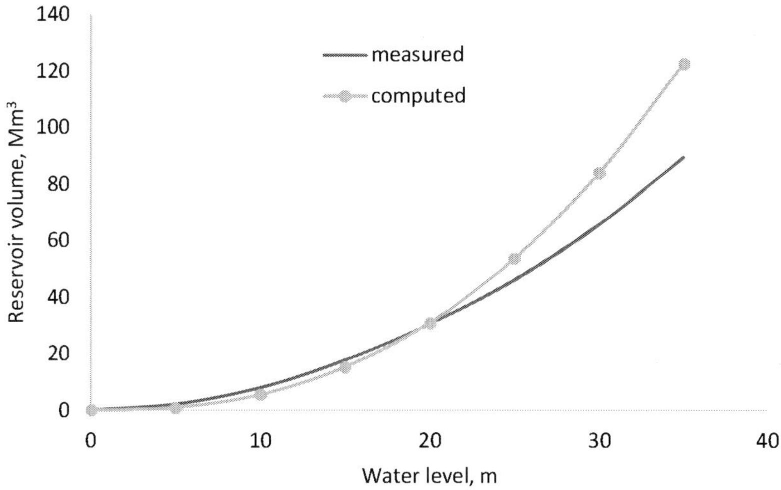

Figure 4.11. Measured and calculated values for comparison.

The relationship between water depth and volume of the reservoir is obtained as:

$$V = Ah^B$$

$$V = 0.01936h^{2.458134}$$

The measured and calculated values of reservoir volume vs elevation using the above relation are presented in Table 4.8 and in Figure 4.11 for comparison.

4.6.5. Applications of Numerical Integration Using Simpson's Rule

Use Simpson's rule to calculate the area under the function over the range of x = [3 6] for the function $f(x) = 5x^2 + 8x + 1$.

Solution: The goal is to solve the integral given below,

$$A = \int_{x_1=3}^{x_2=6} f(x)\,dx$$

Analytical solution,

$$A = \int_3^6 (5x^2 + 8x + 1)dx = \left|\frac{5x^3}{3} + \frac{8x^2}{2} + x\right| = \left(\frac{5\times216}{3} + \frac{8\times36}{2} + 6\right) - \left(\frac{5\times27}{3} + \frac{8\times9}{2} + 3\right)$$
$$= 426$$

Assuming seven ordinates ($n = 7$), the number of intervals that should be an even number is $n - 1 = 6$. Therefore,

$$a = \frac{x_n - x_1}{n-1} = \frac{6-3}{7-1} = 0.5$$

$$x_1 = 3.0,\ x_2 = 3.5,\ x_3 = 4.0,\ x_4 = 4.5,\ x_5 = 5,\ x_6 = 5.5,\ x_7 = 6,$$

$$h_1 = f(x_1) = 5 \times 9 + 8 \times 3 + 1 = 70.$$

Similarly h_1 to h_7 is calculated as below. $h_2 = 90.5$; $h_3 = 113$; $h_4 = 138.25$; $h_5 = 166$; $h_6 = 196.25$; $h_7 = 229$.

Finally, using the Equation (4.19), the area is calculated as:

$$A = \frac{0.5}{3}[70 + 229 + 4(90.5 + 138.25 + 196.25) + 2(113 + 166)] = 426.116$$

Analytical solution is 426. The difference between analytical and numerical solution is 0.116. To achieve more accuracy, if we assume n = 13, we will have 12 intervals and so a = (6 – 3)/12 = 0.25. Table 4.9 shows the resulting ordinates. The corresponding area in this case showed that there is no difference between analytical and numerical solutions.

Table 4.9. Numerical solution for n = 13

n	1	2	3	4	5	6	7	8	9	10	11	12	13
x	3	3.25	3.5	3.75	4	4.25	4.5	4.75	5	5.25	5.5	5.75	6
f(x)	70	79.8	90.3	101	113	125	138	152	166	181	196	212	229

$$A = \frac{0.25}{3}[70 + 229 + 4(79.8 + 101 + 125 + 152 + 181 + 212) + 2(90.3 + 113 + 138 + 166 + 196)] = 426.$$

4.6.6. Determination of Emptying Time of a Reservoir with Constant Inflow

Considering the information given below, determine the emptying time of a reservoir from the elevation of 37 to 25 m.

Given data:

Storage-elevation relationship for reservoir: $V = 8000\,H^{2.7}$, (V is Volume in m^3 and H is height in m).

Diameter of the outlet: $D = 2m$,

Discharge coefficient for outlet: $c = 0.65$ (assumed constant at each elevation)

Constant input discharge: $I = 10.0\ m^3/s$

Outlet level from the reservoir bottom: h = 8 m.

Solution: Using continuity equation dV/dt = I – O in which t is the time, O is the outflow, and I is the inflow. Elevation change in time can be expressed as below:

$$\int_{t_1}^{t_2} dt = \int_{h_1}^{h_2} \frac{dV}{I - O} \tag{4.38}$$

or

$$\int_0^T dt = T = \int_{h2}^{h1} \frac{dV}{O - I} \tag{4.39}$$

We assumed that there is a relationship between volume and water depth of reservoir as $V = aH^b$. If we differentiate this equation, we can express it as:

$$dV = ab(H^{b-1})dh \quad (4.40)$$

Considering points 0 and 1 in Figure 4.12, the velocity at the orifice outlet can be obtained from Bernoulli equation as:

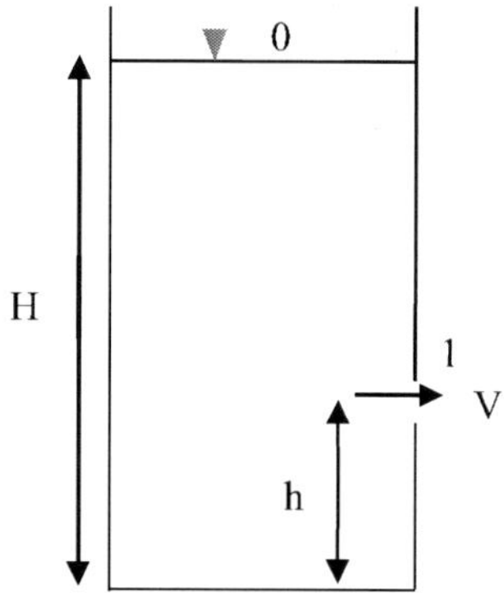

Figure 4.12. Orifice velocity considering points 0 and 1.

$$V = C[2g(H - h)]^{0,5} \quad (4.41)$$

in which, V is the velocity at orifice outlet,
C is the discharge coefficient,
H is the water depth form the bottom of reservoir,
h is the orifice level from the bottom,
g is the gravitational acceleration.
Discharge from outlet is O = VA, so

$$O = CA[2g(H - h)]^{0,5} \quad (4.42)$$

We may construct an equation for emptying time as Equation (4.39)

$$T = \int \frac{abh^{b-1}}{C_0 r^2 (2gh)^{0,5} - I} dh \quad (4.43)$$

or

$$T = \int_{h1}^{h2} F(H)dh \quad (4.44)$$

where

$F(H) = ab(H^{b-1})/\{[CA[2g(H\text{-}h)]^{0.5}]] - I\}$

Data: $a = 8000$; $b = 2.7$; $C = 0.65$; $A = \pi D^2/4 = \pi x2x2/4 = 3.14\ m^2$; H_1 = 25 m; $H_2 = 37$ m; $h = 8$ m; $I = 10\ m^3/s$. Using the given data, we obtain,

$F(H) = ab(H^{b-1})/\{[\ CA[2g(H - h)]^{0.5}\ ]- I\}$

$F(H) = 8000\times 2.7\times H^{1.7}/[0.65\times 3.14\times (19.62)^{0.5}\times (H - 8)^{0.5} - 10]$

$F(H) = 21600\times H^{1.7}/[9.045x(H - 8)^{0.5} -10]$

We can calculate F(H) from the above equation using different water levels as H.

If we assume n = 7, we will have 6 intervals and so a = (37 - 25)/6 = 2. Table 4.10 shows the resulting values of F(H).

Table 4.10. Calculated values for F(H)

n	1	2	3	4	5	6	7
H (m)	25.00	27.00	29.00	31.00	33.00	35.00	37.00
F(H)	188318	199084	210338	221975	233925	246137	258575

Using the Simpson's rule, we obtain the emptying time as,

$T = 2669471\ s = 30.89666\ days \approx 31\ days$

4.6.7. Determination of Emptying Time for a Small Reservoir

As a hydraulic engineer, you are requested to estimate the evacuation time for a small irrigation reservoir with a storage capacity given in the form of $V = ah^b$ where V = volume, h = height. Find the emptying time for a reservoir with the following data.

Given data: a =1000; b = 3; discharge coefficient of orifice: C_0 = 0.63; r = 1.25 m (as a square cross section); h_1 = 50 m; h_2 = 40 m; Inflow rate: I = 25 m^3/s; n = 5.

Solution: A sketch is shown in Figure 4.13 for definition of the problem.

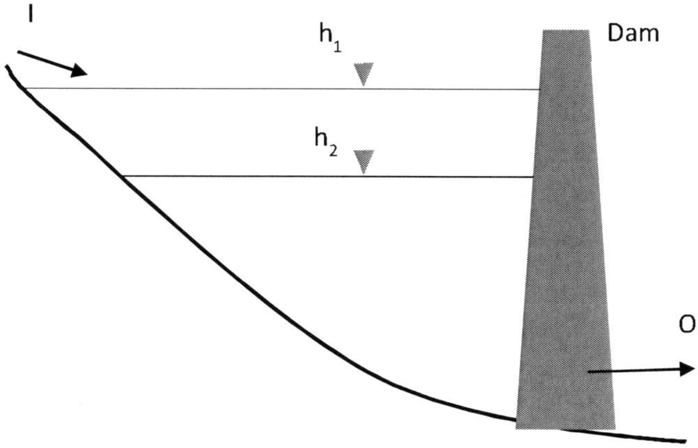

Figure 4.13. Dam and reservoir scheme.

Reservoir emptying time can be calculated from Equation (4.43)

$$T = \int \frac{abh^{b-1}}{C_0 r^2 (2gh)^{0,5} - I} dh$$

Using the data as a =1000; b = 3; C_0= 0.63; r = 1.25m; h_1= 50 m; h_2= 40 m; I = 25; n = 5.

From the previous subtitles, the emptying time,

$$T = \int_{40}^{50} \frac{abh^{b-1}}{C_0 r^2 (2gh)^{0.5} - I} dh$$

$$f(h) = \frac{abh^{b-1}}{C_0 r^2 (2gh)^{0.5} - I} = \frac{100 \times 3 \times h^2}{0.63 \times 1.25^2 (2 \times 9.81 h)^{0.5} - 25}$$

$$f(h) = \frac{300h^2}{4.3586h^{0.5} - 25}$$

$$a = \frac{h_n - h_i}{n-1} = \frac{50-40}{4} = 2.5$$

For $h_1 = 50 \rightarrow f(50) = \frac{3000 \times 50^2}{4.3586 x 50^{0.5} - 25} = 1286111.8$

For $h_2 = 47.5 \rightarrow f(47.5) = 1340118.7$

For $h_3 = 45 \rightarrow f(45) = 1429627.8$

For $h_4 = 42.5 \rightarrow f(42.5) = 1581993.8$

For $h_5 = 40 \rightarrow f(40) = 1862949.8$

Using Simpson's Rule, we can obtain:

$$T = \frac{2.5}{3}[1286111.8 + 1862949.8 + 4(1340118.7 + 1581993.8) + 2 \times 1429627.8]$$

$$T = 14747306 \, \text{sec} = 4096 \text{ hr} \approx 170.7 \text{days}$$

4.6.8. Determination of Emptying Time of a Reservoir with No Inflow

What is the time required to drop the gage height from h_1 to h_2 for the same problem at section 4.6.7 assuming no inflow from upstream watershed? (Hint: use analytical solution).

Given data: a =1000; b = 3; $C_0 = 0.63$; r = 1.25 m (as a square cross section); $h_1 = 50$ m; $h_2 = 40$ m; I = 0; n = 5.

Solution: Analytical Solution for this special case is possible.

When inflow is zero (I = 0), or very small compared to Outflow (O), analytical solution is possible.

Considering Equation (4.40) we write

$$T = \int \frac{abh^{b-1}}{c_0 r^2 (2gh)^{0.5} - I} dh$$

If I = 0, then we get

$$T = \int_{40}^{50} \frac{1000 \times 3xh^2}{0.63 \times (1.25)^2 (19.62h)^{0.5}} dh$$

We can solve analytically the above equation as

$$T = \int_{40}^{50} 688\, h^{1.5}\, dh$$

Substituting the given data in the above equation yields,

$$T = (688/2.5)|h^{2.5}| = 275.21\times(17677.67 - 10119.29)$$

$$T = 2080142 \text{ seconds} \approx 24 \text{ days.}$$

Chapter 5

Finite Difference Methods

5.1. Introduction

One of the first approaches to numerical solution of differential equations is the finite difference method (Morton and Mayers, 2005). This method, which was first used by Euler in the 18th century, is directly applied to the differential form of basic equations. The most important advantage of the finite difference method is its simplicity. Another advantage is its ability to obtain high-order approximations. In this way, more accurate results can be obtained. However, there are some weaknesses in problems in domains with irregular geometry.

The fundamentals concepts of the calculus of finite differences as it applies to hydraulic computations are presented in this chapter. Emphasis is placed on the types of relationship to the differential equation from which they derive.

5.2. Difference Equation

A difference equation is an algebraic equation relating the values of one or more dependent variables at discrete points in the continuous domain of one or more independent variables. The difference equation is the result of the application of a difference scheme to a differential equation. This allows the conversion of an analytical statement of the problem for which a closed-formed solution may be either difficult or impossible to obtain, into an algebraic statement for which a numerical solution can generally be made available.

The numerical properties of the solution to a difference equation are tried to the physical properties of the continuum problem. Differential equations can be either of the ordinary or partial type. The finite difference methods to be discussed here are particularly suited to the solution of partial differential equations with two independent variables. Therefore, this type of equation will be treated here in detail.

5.3. Partial Differential Equations

Differential equations can be classified as ordinary differential equations and partial differential equations as can be seen in Figure 5.1.

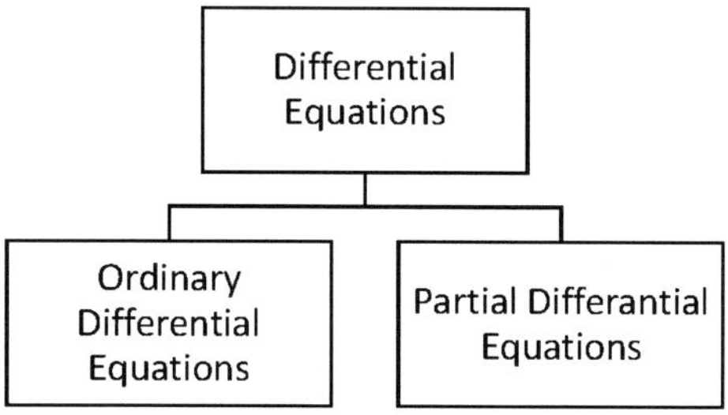

Figure 5.1. Classification of differential equations.

The most common problems in hydraulics are expressible in a single second order partial differential equation. For instance, the two first order mass and momentum conservation equations can be combined into one second order equation.

As summary partial differential equations can be classified as mentioned in Chapter 3.

1. Parabolic Equations
2. Elliptic Equations
3. Hyperbolic Equations.

Elliptic equations are generally related to balance problems or steady state problems. For example, the velocity potential for the steady flow of a non-viscous and incompressible fluid provides the Laplace equation, mathematically. This Laplace equation which is an elliptic partial differential equation must satisfy both in the interior of the domain and its boundaries. The conditions that the dependent variables will meet at the domain boundaries are called boundary conditions. For this, problem of elliptic equation is sometimes called "boundary value" problem or "jury" problem as can be seen in Table 5.1.

Problems involving time as one of the independent variables usually lead to parabolic or hyperbolic equations. In these cases, initial and boundary conditions need to be specified in addition to the equation (Causon and Mingham, 2010).

The application of finite difference methods to elliptic, parabolic or hyperbolic equations follows the same pattern of continuum discretization, scheme selection, boundary value specification and formulation of the algebraic equations linking the values of the dependent variables at neighboring nodes (Hoffman, 2001). However, in contrast to the closed domain associated with the elliptic case, the parabolic and hyperbolic problems exhibit an "open" domain. In this respect, they are also referred to as "mixed initial-boundary value" or "marching" problems, the latter because of the way in which they propagate the solution forward from one-time row to the next in a step-by-step fashion.

The difference between the parabolic equations possess only one solution (+ characteristic curve), while hyperbolic equations possess two solutions (+ and- characteristic curves). Along the characteristic curves, the solution of the partial differential equation reduces to the integration of an ordinary differential equation. This is the basis of the method characteristic which is treated in detail in Chapter 12.

Comparison of the partial differential equations is shown in Table 5.1.

Table 5.1. Comparison of the partial differential equations

Type of equation	Problem	Conditions	Independent variables	Number of Solutions	Solution Domain	Sketch
Elliptic	Jury problem, Boundary value problem	Boundary conditions (B.C.)	x, y	1	Closed	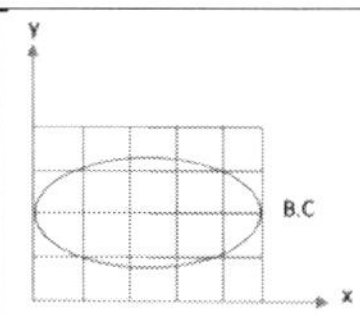
Parabolic	Marching problem*	Boundary condition + Initial condition (B.C., I.C.)	x, t	1	Open	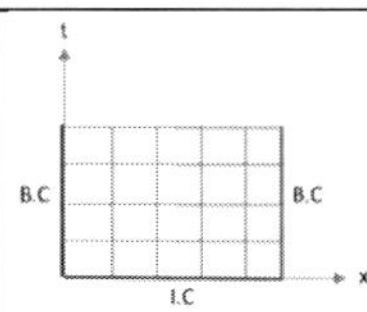
Hyperbolic	Marching problem*	Boundary condition + Initial condition (B.C., I.C.)	x, t	2	Open	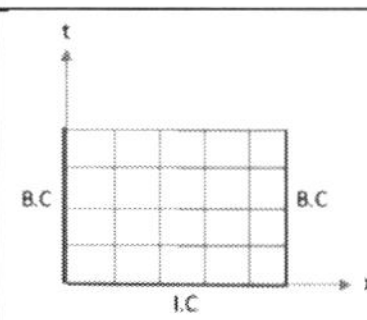

*Mixed initial and boundary value problem.

5.4. Derivation of Some Basic Finite Difference Forms

5.4.1. Taylor Series Expansions

The basic finite difference forms for partial derivatives can be derived from the partial Taylor series expansions. The rectangular mesh used is shown in Figure 5.2. Subscripts (i, j) refer to x, y indices, and superscripts (n) refer to time levels. In a two-dimensional domain, the mesh spacing in the i, j directions is x and Δy. f variable is any function.

The first non-centered derivative forms for $\partial f/\partial x$ can be derived as follows. Here, the continuity of the derivatives is assumed and expands forward in a Taylor series. The superscript (n) has been omitted for ease of writing and clarity.

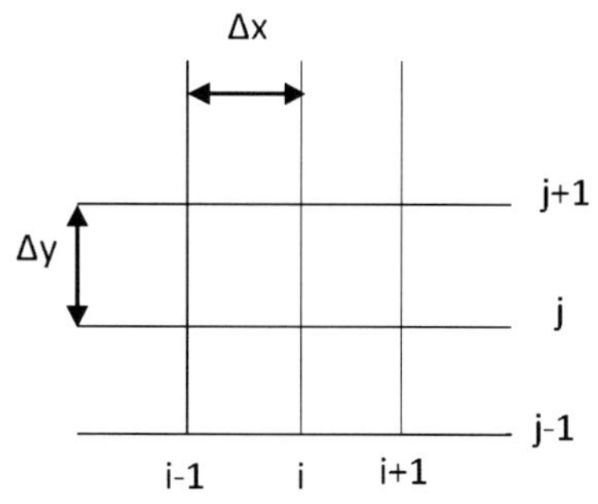

Figure 5.2. Finite difference grid for rectangular geometry.

The same notation, f, can be used for the continuous function f (x, y) and the discrete function f (i, j).

$$f_{i+1,j} = f_{i,j} + \frac{\partial f}{\partial x}\Big|_{i,j}(x_{i+1,j} - x_{i,j}) + \frac{1}{2}\frac{\partial^2 f}{\partial x^2}\Big|_{i,j}(x_{i+1,j} - x_{i,j})^2 + \cdots = f_{i,j} + \frac{\partial f}{\partial x}\Big|_{i,j}\Delta x + \frac{1}{2}\frac{\partial^2 f}{\partial x^2}\Big|_{i,j}\Delta x^2 + \text{HOT} \quad (5.1)$$

where HOT is an abbreviation for higher-order terms.

Solving for $\partial f/\partial x$ gives:

$$\frac{\partial f}{\partial x}\Big|_{i,j} = \frac{f_{i+1,j} - f_{i,j}}{\Delta x} - \frac{1}{2}\frac{\partial^2 f}{\partial x^2}\Big|_{i,j}\Delta x + \text{HOT} = \frac{f_{i+1,j} - f_{i,j}}{\Delta x} + 0(\Delta x) \quad (5.2)$$

where $0(\Delta x)$ is read "terms of order Δx," and refers to additional terms with factors of Δx, Δx^2, Δx^3, etc. We denote any finite-difference analog of $\partial f/\partial x$. Then the forward difference approximation $\partial f/\partial x$ is

$$\frac{\partial f}{\partial x}\bigg|_{i,j} = \frac{f_{i+1,j} - f_{i,j}}{\Delta x} \tag{5.3}$$

This approximation has a truncation error of order Δx; that is, a "first order" accuracy.

By expanding backwards to $f_{i-1,j}$, we obtain another finite-difference analog ($\partial f/\partial x$), this one being the backward-difference approximation can be expressed as,

$$\frac{\partial f}{\partial x}\bigg|_{ij} = \frac{f_{i,j} - f_{i-1,j}}{\Delta x} \tag{5.4}$$

This approximation is also first-order accurate. The centered difference approximation $\partial f/\partial x$ is obtained by subtracting the forward and backward expansions as follows,

$$f_{i+1,j} = f_{i,j} + \frac{\partial f}{\partial x}\bigg|_{i,j} \Delta x + \frac{1}{2}\frac{\partial^2 f}{\partial x^2}\bigg|_{i,j}\Delta x^2 + \frac{1}{6}\frac{\partial^3 f}{\partial x^3}\bigg|_{i,j}\Delta x^3 + \frac{1}{24}\frac{\partial^4 f}{\partial x^4}\bigg|_{i,j}\Delta x^4 + 0(\Delta x^5) \tag{5.5}$$

$$f_{i-1,j} = f_{i,j} - \frac{\partial f}{\partial x}\bigg|_{i,j} \Delta x + \frac{1}{2}\frac{\partial^2 f}{\partial x^2}\bigg|_{i,j}\Delta x^2 - \frac{1}{6}\frac{\partial^3 f}{\partial x^3}\bigg|_{i,j}\Delta x^3 + \frac{1}{24}\frac{\partial^4 f}{\partial x^4}\bigg|_{i,j}\Delta x^4 + 0(\Delta x^5) \tag{5.6}$$

Subtracting Equation (5.9) from Equation (5.10) gives

$$f_{i+1,j} - f_{i-1,j} = 2\frac{\partial f}{\partial x}\bigg|_{i,j} \Delta x + \frac{1}{3}\frac{\partial^3 f}{\partial x^3}\bigg|_{i,j}\Delta x^3 + \text{HOT} \tag{5.7}$$

Solving for $\partial f/\partial x$ gives

$$\frac{\partial f}{\partial x}\bigg|_{i,j} = \frac{f_{i+1,j} - f_{i-1,j}}{2\Delta x} - \frac{1}{6}\frac{\partial^3 f}{\partial x^3}\bigg|_{i,j}\Delta x^2 + \text{HOT} = \frac{f_{i+1,j} - f_{i-1,j}}{2\Delta x} + 0(\Delta x^2) \tag{5.8}$$

The centered difference form of $\partial f/\partial x$ is then

$$\frac{\partial f}{\partial x}\bigg|_{i,j} = \frac{f_{i+1,j} - f_{i-1,j}}{2\Delta x} \tag{5.9}$$

This approximation has a truncation error of order Δx^2; that is, a "second order" accuracy. Analogous expressions follow immediately for y- and t-derivatives. For example, the centered difference analog of $\partial f/\partial t$ is

$$\frac{\partial f}{\partial t}\bigg|_{i,j}^{n} = \frac{f_{i,j}^{n+1} - f_{i,j}^{n-1}}{2\Delta t} \tag{5.10}$$

We derive a centered difference analog of $(\partial^2 f/\partial x^2)$ by adding the Equation (5.9) and Equation (5.10).

$$f_{i+1,j} + f_{i-1,j} = 2f_{i,j} + \frac{\partial^2 f}{\partial x^2}\bigg|_{ij} \Delta x^2 + \frac{1}{12}\frac{\partial^4 f}{\partial x^4}\bigg|_{ij} \Delta x^4 + \text{HOT} \tag{5.11}$$

Solving for $(\partial^2 f/\partial x^2)$ gives

$$\frac{\partial^2 f}{\partial x^2}\bigg|_{i,j} = \frac{f_{i+1,j} + f_{i-1,j} - 2f_{i,j}}{\Delta x^2} + 0(\Delta x^2) \tag{5.12}$$

The centered difference form of $\frac{\partial^2 f}{\partial x^2}$ is then

$$\frac{\partial^2 f}{\partial x^2}\bigg|_{i,j} = \frac{f_{i+1,j} + f_{i-1,j} - 2f_{i,j}}{\Delta x^2} \tag{5.13}$$

which is second-order accurate. Some finite difference formulas are summarized in section 5.6.

Similarly, we can derive the following expression for $(\partial^2 f/\partial x\partial y)$ as,

$$\frac{\partial^2 f}{\partial x\, \partial y}\bigg|_{i,j} = \frac{f_{i+1,j+1} - f_{i+1,j-1} - f_{i-1,j+1} - f_{i-1,j-1}}{4\Delta x\Delta y} \tag{5.14}$$

A casual carrying of the order of truncation terms in the first method would indicate only first-order accuracy. But, by carefully carrying high-order terms in the second method, one can demonstrate second-order accuracy, with truncation error $E = 0(\Delta x^2 + \Delta y^2)$. In Equation (5.18), $\partial^2 f/(\partial x\partial y)$ obeys the continuum operative rule as, $\partial^2 f/(\partial x\partial y) = \partial^2 f/(\partial y\partial x)$.

Combinations of such finite-difference expressions for partial derivatives alone may be put together to form finite-difference expressions for the partial differential equations.

For example, the Laplace equation becomes

$$\frac{\partial^2 f}{\partial x^2}+\frac{\partial^2 f}{\partial y^2}=0 \rightarrow \frac{f_{i+1,j}+f_{i-1,j}-2f_{i,j}}{\Delta x^2}+\frac{f_{i,j+1}+f_{i,j-1}-2f_{i,j}}{\Delta y^2}=0 \tag{5.15}$$

$$f_{i+1,j}+f_{i-1,j}+\beta^2\left(f_{i,j+1}+f_{i,j-1}\right)-2(1+\beta^2)f_{i,j}=0 \tag{5.16}$$

where β is the mesh aspect ratio, $\beta = (\Delta x/\Delta y)$. This is commonly referred to as the "5-point stencil." For $\beta = 1$, we obtain the well-known equation.

$$f_{i+1,j}+f_{i-1,j}+\left(f_{i,j+1}+f_{i,j-1}\right)=4f_{i,j} \tag{5.17}$$

which says that the value $f_{i,j}$ is the average of f at the four neighboring points. These forms are indicated schematically in Figure 5.3.

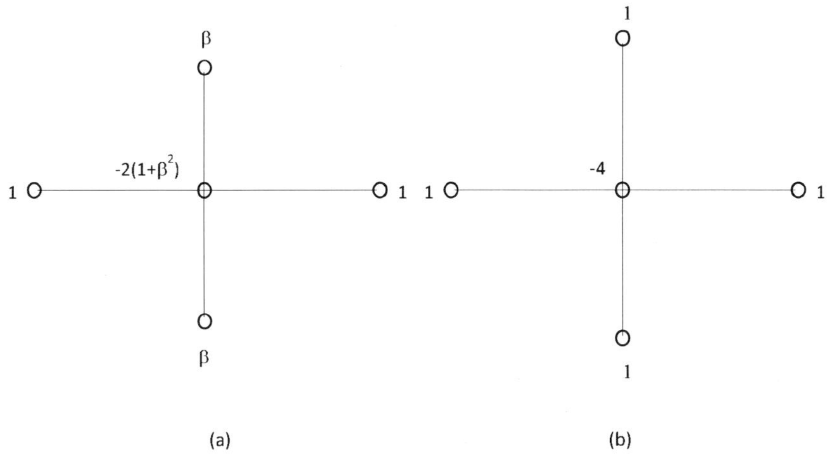

Figure 5.3. Schematic representation of the 5-point Laplace equation, a) for Arbitrary β and b) $\beta = 1$.

A simple expression for the linear model Equation (5.3) with c = 1, using second-order-accurate expressions for both space and time derivatives would be as

$$\frac{u_i^{n+1} - u_i^{n-1}}{2\Delta t} = \frac{u_{i+1}^n - u_{i-1}^n}{2\Delta x} + \alpha \frac{u_{i+1}^n + u_{i-1}^n - 2u_i^n}{\Delta x^2} \tag{5.18}$$

from which one could solve explicitly for u_1^{n+1} in terms of previous values. This method is, in fact, unacceptable. For all $\alpha > 0$ and for all time steps $\Delta t > 0$, the method is numerically unstable, meaning that chaotic solutions with no relation to the continuum solution will be generated. This behavior emphasizes the difference between an accurate finite-difference analog for a derivative, and accuracy for a differential equation.

If, instead of all centered differences, we use forward differences in time, we obtain an analog of the linear model equation (5.3) which is second-order accurate in space, and only first order in time,

$$\frac{u_i^{n+1} - u_i^n}{\Delta t} = \frac{u_{i+1}^n - u_{i-1}^n}{2\Delta x} + \alpha \frac{u_{i+1}^n + u_{i-1}^n - 2u_i^n}{\Delta x^2} \tag{5.19}$$

We will refer to this method, using forward-time and centered-space differences, as the FTCS method. We will show later that this expression is stable, at least for some conditions, depending on Δt, u, α and Δx. But before we cover stability analyses, we consider a few other aspects of finite-difference equations.

5.4.2. Polynomial Fitting

Another method to obtain finite difference expressions is to fit a free parameter analytic function to mesh point values. This is a common method of deriving derivatives from experimental data. Ideally, the functional form of fit is determined by an approximate analytical solution. Polynomials are more commonly used for this purpose. The method is shown with a parabolic fit.

Given data f at (i-1), (i), and (i+1), we assume a parabolic form

$$f(x) = a + bx + cx^2 \tag{5.20}$$

For convenience, the origin x = 0 is translated to location (i). Then equation (5.20) written at (i-1), (i), and (i+1) gives respectively,

$$f_{i-1} = a - b\Delta x + c\Delta x^2 \tag{5.21}$$

$$f_i = a \tag{5.22}$$

$$f_{i+1} = a + b\Delta x + c\Delta x^2 \tag{5.23}$$

Adding the f_{i-1} and f_{i+1} , equations give

$$c = \frac{f_{i+1} + f_{i-1} - 2f_i}{2\Delta x^2} \tag{5.24}$$

Solving for b gives

$$b = \frac{f_{i+1} - f_{i-1}}{2\Delta x} \tag{5.25}$$

Differentiating equation (5.20) and evaluating at (i) gives

$$\frac{\delta f}{\delta x}\bigg|_i = [b + 2cx]_{x=0} = b = \frac{f_{i+1} - f_{i-1}}{2\Delta x} \tag{5.26}$$

and

$$\frac{\delta^2 f}{\delta x^2}\bigg|_i = 2c = \frac{f_{i+1} + f_{i-1} - 2f_i}{\Delta x^2} \tag{5.27}$$

These two equations, with Equation (5.24) and Equation (5.25), give identically the same results as those of the centered second-order Taylor series expansions, Equation (5.9) and Equation (5.13), respectively.

If a first-order polynomial is assumed, $f = a + bx$, then the forward or backward difference for $(\partial f/\partial x)$ is obtained, depending on whether f_i and f_{i+1} are used to fit a 2nd b, or f_i and f_{i-1}. Obviously, no expression can be obtained for $(\partial^2 f/\partial x^2)$ by fitting f with a straight line.

5.5. Types of Finite Difference Schemes

5.5.1. Basic Forms

The solution by a finite difference method requires that the domain of the independent variables be discretized into a finite number of grid points. The set of grid points encompassed within the domain is called the grid system.

The pattern and spacing of grid points determine the associated numerical properties of the finite difference scheme, namely stability, consistency and convergence. These properties vary with the nature of the problem and with the structure of the numerical scheme being used to solve it.

A difference operator can be either: (a) forward, (b) backward or (c) central as derived in (5.4.1).

5.5.2. Geometric Interpretation of Basic Finite Difference Forms

Partial or ordinary first order derivative can be defined the slope of function at a given point. Mathematically this can be written as

$$\frac{df}{dx} = m = \text{slope} \tag{5.28}$$

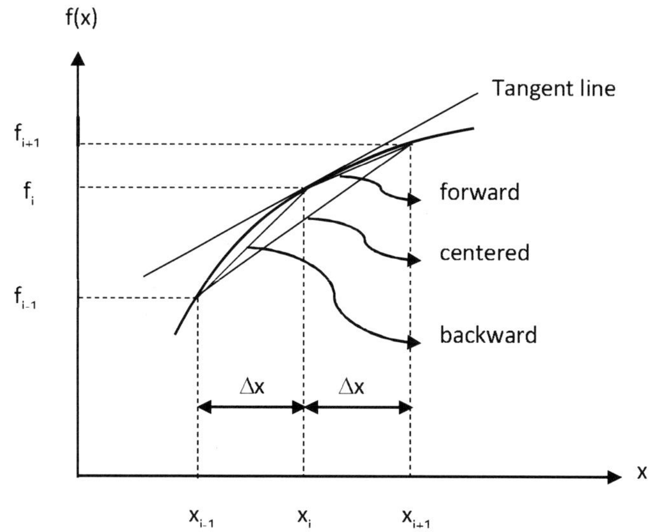

Figure 5.4. Geometric interpretation of basic finite difference forms.

Different geometric interpretations of the first-order finite difference approximation related to forward, backward, and central difference approximation can be defined as shown in Figure 5.4;

a. For forward finite difference:

$$\left.\frac{\partial f}{\partial x}\right|_i = \frac{f_{i+1} - f_i}{\Delta x} \tag{5.29}$$

b. For backward finite difference:

$$\left.\frac{\partial f}{\partial x}\right|_i = \frac{f_i - f_{i-1}}{\Delta x} \tag{5.30}$$

c. For centered finite difference:

$$\left.\frac{\partial f}{\partial x}\right|_i = \frac{f_{i+1} - f_{i-1}}{2\Delta x} \tag{5.31}$$

5.5.3. Explicit and Implicit Finite Difference Schemes

Finite difference schemes can be either (a) explicit or (b) implicit. An explicit scheme is that in which the value of the unknown dependent variable at any point can be calculated exclusively in terms of known neighboring values. In such a case, the unknown value can be related to the known values by means of an algebraic equation, and solved directly (i.e, explicitly). For instance, in Figure 5.5, the value of $_i^{n+1}$ can be related to the values of ϕ at the known level n (λ).

$$\phi_i^{n+1} = f(\phi_{i-1}^n, \phi_i^n, \phi_{i+1}^n) \tag{5.32}$$

An implicit scheme is that in which the values of the unknown variables occur in pairs or triads in the difference equation. The set of difference equations constitute a set of simultaneous algebraic equations, and the solution for any one point can only be obtained by carrying out the simultaneous solution of all points in the domain of the dependent variable. In such a case, the scheme is said to be implicit, because in contrast to the explicit type, no direct point-by-point solution is possible. For instance, in Figure 5.6, the value of ϕ at the unknown level n+1 and ϕ_{i+1}^{n+1} can be used to express a space difference operator, while ϕ_i^{n+1} and ϕ_i^n can be used to express a time difference operator. The unknown values occur in triads, and the solution is carried out by matrix inversion producers. In implicit schemes no direct point by point solution is possible.

$$\phi_i^{n+1} = f[\phi_{i-1}^{n+1}, \phi_i^n, \phi_{i+1}^{n+1}] \tag{5.33}$$

In this formula, n+1 shows the function in the unknown level.

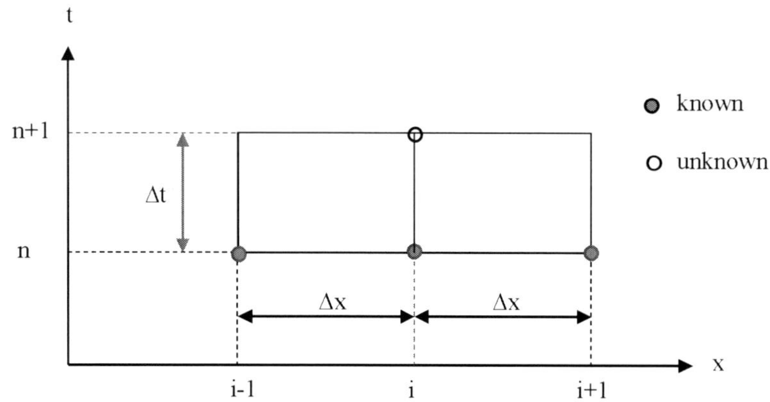

Figure 5.5. Explicit scheme.

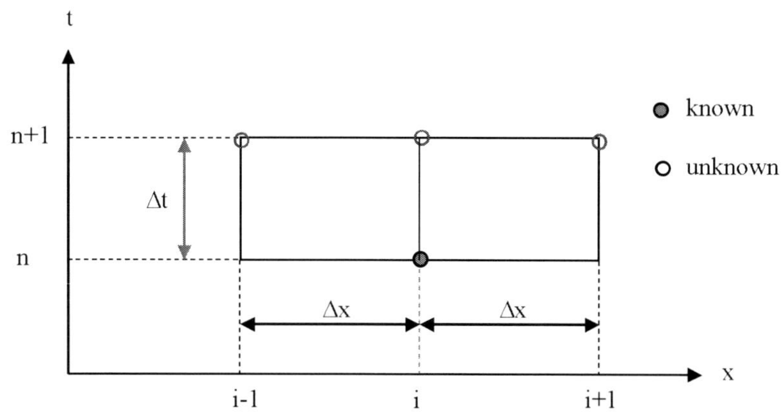

Figure 5.6. Implicit scheme.

The finite difference approach is popular due to its ease of formulation and execution. Numerous numerical schemes are established based on different methods of discretization. Some of them are given below.

A. Explicit finite difference: For a function f at point (i, n) can be written as (see Figure 5.6),

$$\text{Forward: } \frac{df}{dx}\Big|_i^n = \frac{f_{i+1}^n - f_i^n}{\Delta x} \tag{5.34}$$

$$\text{Backward: } \frac{df}{dx}\Big|_i^n = \frac{f_i^n - f_{i-1}^n}{\Delta x} \tag{5.35}$$

$$\text{Centered: } \frac{df}{dx}\Big|_i^n = \frac{f_{i+1}^n - f_{i-1}^n}{2\Delta x} \tag{5.36}$$

B. Implicit finite difference: For a function f at point (i, n) can be written as (see Figure 5.6),

$$\text{Backward: } \frac{df}{dx}\Big|_i^n = \frac{f_i^{n+1} - f_{i-1}^{n+1}}{\Delta x} \tag{5.37}$$

$$\text{Forward: } \frac{df}{dx}\Big|_i^n = \frac{f_{i+1}^{n+1} - f_i^{n+1}}{\Delta x} \tag{5.38}$$

$$\text{Central: } \frac{df}{dx}\Big|_i^n = \frac{f_{i+1}^{n+1} - f_{i-1}^{n+1}}{2\Delta x} \tag{5.39}$$

5.5.4. Comparison of Explicit and Implicit Schemes

Explicit schemes are simple to formulate but are usually subject to a necessary condition for their stability. This effectively sets an upper limit to the size of the time interval Δt that can be used in practice.

Implicit schemes, on the other hand, are not subject to a necessary condition for their stability, thereby allowing the use of larger time intervals. However, the size of the time interval cannot be set arbitrarily. Care should be taken to ensure that the continuum function is being properly discretized. Therefore, the selection of the time interval is primarily dictated by accuracy considerations.

Not all explicit schemes are subject to a stability limitation; for certain types of continuum problems, unconditionally stable explicit schemes can be constructed. Likewise, not all implicit schemes are free from stability problems. Instability due to round-off error or nonlinear effects sometimes plague implicit numerical computations.

The advantage of being able to use a comparably larger time interval in implicit models is somewhat offset by the additional complications involved in their formulation (including the need to solve sets of simultaneous equations). Nevertheless, the accumulated experience with hydraulic applications of numerical modeling has shown that the implicit model is usually more efficient in the use of computational resources that its explicit counterpart. Unfortunately, no such statement can be for the general case.

Comparison of Explicit and Implicit Schemes is made in Table 5.2. In this table Δt is the time interval, and Δt_c is the critical time interval. The explicit scheme is not stable (but it can be accurate and convergent) while the implicit scheme is stable (but not accurate). Figure 5.7 and Figure 5.8 show example solutions for unstable and stable schemes, respectively.

Table 5.2. Comparison of Explicit and Implicit Schemes

Scheme	Stability	Convergence	Programming
Explicit	Limited to $\Delta t \leq \Delta t_c$	Improve near Δt_c	More Simple
Implicit	Unconditionally stable	Continuous being progress	More difficult (Include (matrix or iterations)

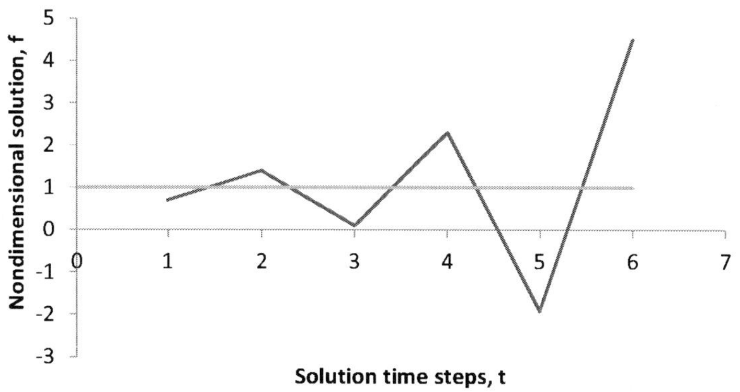

Figure 5.7. Stability description for unstable schemes.

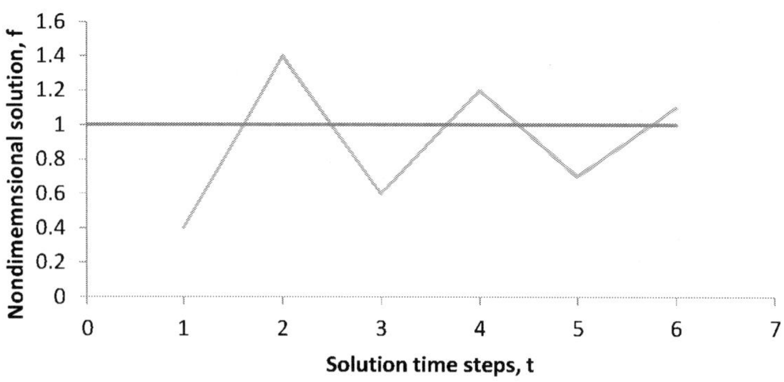

Figure 5.8. Stability description for stable schemes.

5.6. Finite Difference Schemes

We can write different finite difference formulas considering any function f at point i, as can be seen in Figure 5.9.

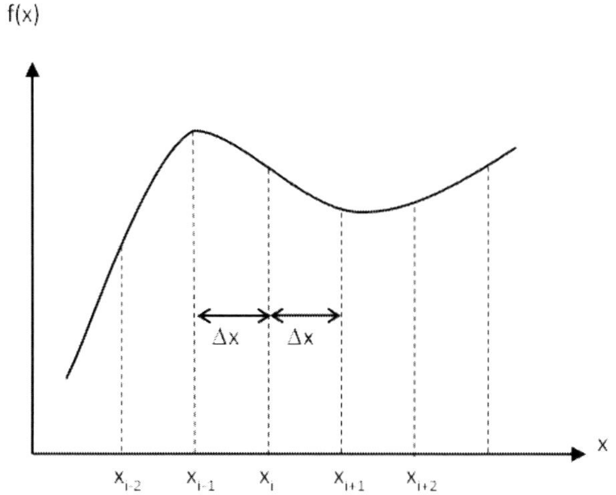

Figure 5.9. Finite difference discretization for any function f for Δx = h.

5.6.1. Finite Difference Formulas for First Derivatives

Finite difference formula for first order derivatives at point i are shown in Table 5.3. In this table, h is the space interval as Δx (see Figure 5.9), and O(h) indicates the order of truncation error.

Table 5.3. Finite difference formulas for first order derivatives

Finite Difference Form	Formula	Truncation Error
Two-Point Backward Difference	$f'(x_i) = \frac{f(x_i)-f(x_{i-1})}{h}$	O(h)
Three-Point Backward Difference	$f'(x_i) = \frac{f(x_{i-2})-4f(x_{i-1})+3f(x_i)}{2h}$	$O(h^2)$
Two-Point Central Difference	$f'(x_i) = \frac{f(x_{i+1})-f(x_{i-1})}{2h}$	$O(h^2)$
Four-Point Central Difference	$f'(x_i) = \frac{f'(x_{i-2})-8f(x_{i-1})+8f(x_{i+1})-f(x_{i+2})}{12h}$	$O(h^4)$

5.6.2. Finite Difference Formulas for Second Derivatives

Finite difference formula for second order derivatives at point i are shown in Table 5.4. In this table, O(h) indicates the order of accuracy.

Table 5.4. Finite difference formulas for second order derivatives

Finite Difference Form	Formula	Truncation Error
Three-Point Forward Difference	$f''(x_i) = \frac{f(x_i)-2f(x_{i+1})+f(x_{i+2})}{h^2}$	O(h)
Four-Point Forward Difference	$f''(x_i) = \frac{2f(x_i)-5f(x_{i+1})+4f(x_{i+2})-f(x_{i+3})}{h^2}$	$O(h^2)$
Three-Point Backward Difference	$f''(x_i) = \frac{f(x_{i-2})-2f(x_{i-1})+f(x_i)}{h^2}$	O(h)
Four-Point Backward Difference	$f''(x_i) = \frac{-f(x_{i-3})+4f(x_{i-2})-5f(x_{i-1})+2f(x_i)}{h^2}$	$O(h^2)$
Three-Point Central Difference	$f''(x_i) = \frac{f(x_{i-1})-2f(x_i)+f(x_{i+1})}{h^2}$	$O(h^2)$
Five-Point Central Difference	$f''(x_i) = \frac{-f(x_{i-2})+16f(x_{i-1})-30f(x_i)+16f(x_{i+1})-f(x_{i+2})}{12h^2}$	$O(h^4)$

5.6.3. Finite Difference Formulas for Third Derivatives

Finite difference formulas for third order derivatives at point i are shown in Table 5.5. In this table, O(h) indicates the order of truncation error.

Table 5.5. Finite difference formulas for third derivatives

Finite Difference Form	Formula	Truncation Error
Four-Point Forward Difference	$f'''(x_i) = \frac{-f(x_i)+3f(x_{i+1})-3f(x_{i+2})+f(x_{i+3})}{h^3}$	O(h)
Five-Point Forward Difference	$f'''(x_i) = \frac{-5f(x_i)+18f(x_{i+1})-24f(x_{i+2})+14f(x_{i+3})-3f(x_{i-4})}{2h^3}$	$O(h^2)$
Four-Point Backward Difference	$f'''(x_i) = \frac{-f(x_{i-3})+3f(x_{i-2})-3f(x_{i-1})+f(x_i)}{h^3}$	O(h)
Five-Point Backward Difference	$f'''(x_i) = \frac{3f(x_{i-4})-14f(x_{i-3})+24f(x_{i-2})-18f(x_{i-1})+5f(x_i)}{2h^3}$	$O(h^2)$
Four-Point Central Difference	$f'''(x_i) = \frac{-f(x_{i-2})+2f(x_{i-1})-2f(x_{i+1})+f(x_{i+2})}{2h^3}$	$O(h^2)$
Six-Point Central Difference	$f'''(x_i) = \frac{f(x_{i-3})-8f(x_{i-2})+13f(x_{i-1})-13f(x_{i+1})+8f(x_{i+2})-f(x_{i+3})}{8h^3}$	$O(h^4)$

5.6.4. Finite Difference Formulas for Fourth Derivatives

Finite difference formulas for fourth order derivatives at point i are shown in Table 5.6. In this table, O(h) indicates the order of accuracy.

Table 5.6. Finite difference formulas for fourth derivatives

Finite difference Form	Formula	Truncation Error
Five-Point Forward Difference	$f^{iv}(x_i) = \frac{f(x_i)-4f(x_{i+1})+6f(x_{i+2})-4f(x_{i+3})+f(x_{i+4})}{h^4}$	$O(h)$
Six-Point Forward Difference	$f^{iv}(x_i) = \frac{3f(x_i)-14f(x_{i+1})+26f(x_{i+2})-24f(x_{i+3})+11f(x_{i+4})-2f(x_{i+5})}{h^4}$	$O(h^2)$
Five-Point Backward Difference	$f^{iv}(x_i) = \frac{f(x_{i-4})-4f(x_{i-3})+6f(x_{i-2})-4f(x_{i-1})+f(x_i)}{h^4}$	$O(h)$
Six-Point Backward Difference	$f^{iv}(x_i) = \frac{-2f(x_{i-5})+11f(x_{i-4})-24f(x_{i-3})+26f(x_{i-2})-14f(x_{i-1})+3f(x_i)}{h^4}$	$O(h^2)$
Five-Point Central Difference	$f^{iv}(x_i) = \frac{f(x_{i-2})-4f(x_{i-1})+6f(x_i)-4f(x_{i+1})+f(x_{i+2})}{h^4}$	$O(h^2)$
Seven-Point Central Difference	$f^{iv}(x_i) = \frac{f(x_{i-3})+12f(x_{i-2})-39f(x_{i-1})+56f(x_i)+39f(x_{i+2})-f(x_{i-3})}{6h^4}$	$O(h^4)$

5.7. Finite Difference Formulations for Some Equations

5.7.1. Pure Diffusion Equation

The diffusion equation given in Equation (5.40) is a parabolic equation. Since the right hand side of the equation contains a second order partial derivative, centered methods should be used. For the left side of this equation, forward or centered techniques can be employed.

$$\frac{\partial C}{\partial t} = D\frac{\partial^2 C}{\partial x^2} \tag{5.40}$$

Forward Time/Centered Space Scheme: If we use forward time for the first order differential and centered space difference scheme for the second order differential, Equation (5.44) can be transformed into:

$$\frac{C_j^{n+1} - C_j^n}{\Delta t} = D\frac{C_{j+1}^n - 2C_j^n + C_{j-1}^n}{\Delta x^2} \tag{5.41}$$

or

$$C_j^{n+1} = C_j^n + D\frac{\Delta t}{\Delta x^2}(C_{j+1}^n - 2C_j^n + C_{j-1}^n) \tag{5.42}$$

5.7.2. Pure Advection Equation

Let us consider the pure advection equation,

$$\frac{\partial C}{\partial t} + u\frac{\partial C}{\partial x} = 0 \tag{5.43}$$

Forward Time/centered Space Scheme: Equation (5.47) can be represented using the forward time and centered space difference approach as

$$\frac{C_j^{n+1} - C_j^n}{\Delta t} + u\frac{C_{j+1}^n - C_{j-1}^n}{2\Delta x} = 0 \tag{5.44}$$

or,

$$C_j^{n+1} = \frac{\gamma}{2}C_{j-1}^n + C_j^n - \frac{\gamma}{2}C_{j+1}^n \tag{5.45}$$

where $\gamma = u\Delta t/\Delta x$ is known as the Courant number is. The use of centered space discretization for the pure advection equation leads in an unconditionally unstable scheme, as will be shown via the Fourier analysis technique.

Forward Time/backward Space Scheme: Equation (5.47) can be expressed as if we adopt a forward time and backward space difference approach.

$$\frac{C_j^{n+1} - C_j^n}{\Delta t} + u\frac{C_j^n - C_{j-1}^n}{\Delta x} = 0 \tag{5.46}$$

or

$$C_j^{n+1} = \gamma C_{j-1}^n + (1-\gamma)C_j^n \tag{5.47}$$

where, γ was defined as $u\Delta t/\Delta x$.

5.7.3. Numerical vs Physical diffusion

Equation 5.47, which deals with pure advection, has an analytical solution that is given as C(x, t)=C_0 (x -ut). According to the analytical solution, a given initial concentration will move with a velocity of u without dissipating or diffusing. The distribution will keep its current form and peak. However, in a numerical solution the peak may be significantly diffused. This resulted from the diffusion brought on by the first-order numerical method.

5.7.4. Truncation Error and Round-off Error

Truncation error can be defined as the difference between the analytical solution of a function and exact solution of the scheme. Round-off error is defined as the difference between the exact solution of the scheme and numerical solution. Total error is defined as the difference between analytical solution and numerical solution.as seen in Figure 5.10.

In the finite difference approximation of the partial derivatives, the space and time intervals, Δx and Δt, of the grid can be selected small and the truncation error decreases. However, choosing smaller intervals requires more computations. Therefore, rounding errors increase in the solution since more computations are made for the same domain and total computation time. The total error in the calculation is a function of this grid size as seen in Figure 5.10. In this figure, it is assumed that the calculation is made for the same domain and total time.

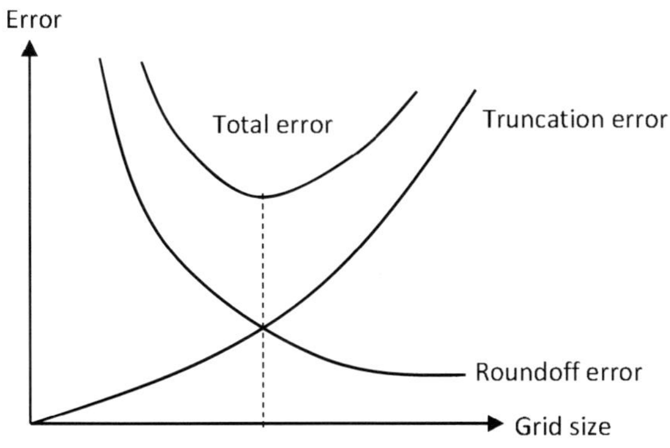

Figure 5.10. Sketch of errors as function of the grid size for a finite difference calculation.

5.8. A Numerical Scheme Analysis

A numerical system can be studied numerically as well as analytically. If an analytical solution is available, solutions from a numerical method can be compared with it. A numerical scheme can be examined analytically using (1) the Taylor series method and (2) the Fourier series method. To evaluate a scheme's consistency, stability, convergence, and degree of precision, analysis is required. Consistency is the condition in the numerical scheme. If the scheme should tend to differential equation when steps in time and space tend to zero, the scheme is consistent. Stability is the condition in numerical solution. All errors should remain limited as the iteration process progresses. For finite values of time and space, the error should be limited when the number of time steps is $n \rightarrow 0$. Convergence is the condition in the numerical solution. If steps in time and space tend to zero, the numerical solution should point towards the exact solution of the mathematical model, and the scheme is convergent. Consistency plus stability equals to convergence.

In time-dependent problems, a finite difference scheme is stable if errors in a time step of the computation do not cause errors to be magnified while calculations are in progress.

The Fourier analysis method provides the stability bound and the quantitative accuracy assessment, while the Taylor series analysis method provides the consistency and order or accuracy (Ali, 2014).

5.8.1. Taylor Series Analysis

For backward finite difference scheme, we consider the forward in time and backward finite difference scheme for Equation (5.47). We can obtain Equation (5.47) as $C_j^{n+1} = \gamma C_{j-1}{}^n + (1-\gamma)C_j^n$.

C_j^{n+1} can be represented using the Taylor series expansion as

$$C_j^{n+1} = C_j^n + \frac{\partial C}{\partial t}\Delta t + \frac{1}{2}\frac{\partial^2 C}{\partial t^2}\Delta t^2 + \frac{1}{6}\frac{\partial^3 C}{\partial t^3}\Delta t^3 + 0(\Delta t^4) \tag{5.48}$$

and $C_{j-1}{}^n$ can be written as:

$$C_{j-1}^n = C_j^n - \frac{\partial C}{\partial x}\Delta x + \frac{1}{2}\frac{\partial^2 C}{\partial x^2}\Delta x^2 - \frac{1}{6}\frac{\partial^3 C}{\partial x^3}\Delta x^3 + 0(\Delta x^4) \tag{5.49}$$

Using Equation (5.52) to Equation (5.53) we can rewrite Equation (5.51) as

$$C_j^n + \frac{\partial C}{\partial t}\Delta t + \frac{1}{2}\frac{\partial^2 C}{\partial t^2}\Delta t^2 + \frac{1}{6}\frac{\partial^3 C}{\partial t^3}\Delta t^3 + 0(\Delta t^4) = \gamma[C_j^n - \frac{\partial C}{\partial x}\Delta x + \frac{1}{2}\frac{\partial^2 C}{\partial x^2}\Delta x^2 + \frac{1}{6}\frac{\partial^3 C}{\partial x^3}\Delta x^3 + 0(\Delta x^4)] + (1-\gamma)C_j^n \quad (5.50)$$

or

$$\frac{\partial C}{\partial t}\Delta t + \frac{1}{2}\frac{\partial^2 C}{\partial t^2}\Delta t^2 + \frac{1}{6}\frac{\partial^3 C}{\partial t^3}\Delta t^3 + 0(\Delta t^4) = u\frac{\Delta t}{\Delta x}[-\frac{\partial C}{\partial x}\Delta x + \frac{1}{2}\frac{\partial^2 C}{\partial x^2}\Delta x^2 - \frac{1}{6}\frac{\partial^3 C}{\partial x^3}\Delta x^3 + 0(\Delta x^4)] \quad (5.51)$$

divide both sides with Δt and make the necessary arrangements,

$$\frac{\partial C}{\partial t} + u\frac{\partial C}{\partial x} = \left[-\frac{1}{2}\frac{\partial^2 C}{\partial t^2}\Delta t + u\frac{1}{2}\frac{\partial^2 C}{\partial x^2}\Delta x\right] + \left[-\frac{1}{6}\frac{\partial^3 C}{\partial t^3}\Delta t^2 - u\frac{1}{6}\frac{\partial^3 C}{\partial x^3}\Delta x^2\right] + 0(\Delta t^3) + 0(\Delta x^3)] \quad (5.52)$$

The left hand side of the equation gives the original equation, $\partial C/\partial t + u\partial C/\partial x = 0$.

Therefore, the error terms are: $\left[-\frac{1}{2}\frac{\partial^2 C}{\partial t^2}\Delta t + u\frac{1}{2}\frac{\partial^2 C}{\partial x^2}\Delta x\right] + \left[-\frac{1}{6}\frac{\partial^3 C}{\partial t^3}\Delta t^2 - u\frac{1}{6}\frac{\partial^3 C}{\partial x^3}\Delta x^2\right] +..0(\Delta t^3) +..0(\Delta x^3)]$.

The dominant terms in the overall error are: $\left[-\frac{1}{2}\frac{\partial^2 C}{\partial t^2}\Delta t + u\frac{1}{2}\frac{\partial^2 C}{\partial x^2}\Delta x\right]$.

The dominant term indicates that the backward finite difference scheme is first order in space and time. Furthermore, the errors are second order partial differential terms, implying that they are diffusive in origin. First-order numerical schemes are hence diffusive (Ali, 2014).

5.8.2. Fourier Analysis

Another analytical method for assessing a numerical scheme is Fourier analysis. In computational hydraulics problems, it is frequently employed. Any linear function can be written as the sum of a variety of trigonometric functions with various amplitudes. In that case, the Fourier transform equation can be used to calculate the amplitudes. The following is a mathematical representation of a function f(x) using the Fourier series (Ali, 2014),

$$f(x) = a_0 + \sum_{k=1}^{\infty}(a_n coskx + b_n sinkx) \quad (5.53)$$

where k = 2π/L; L = wave length; k = wave number; and a_0, a_k, b_k = amplitude. a_o , a_k and b_k can be determined using Fourier transform such as

$$a_0 = (1/\pi)\int_{-\pi}^{\pi} f(x)\, dx \quad (5.54)$$

$$a_k = (1/\pi)\int_{-\pi}^{\pi} f(x)\, coskxdx \quad (5.55)$$

and,

$$b_k = (1/\pi)\int_{-\pi}^{\pi} f(x)\, sinkxdx \quad (5.56)$$

Wave number (k):

The number of waves in a given distance, which is usually two. In other words, it denotes the frequency of a wave length of L, which is equal to 2π /L. The higher the wave number, the smaller the wavelength, and vice versa.

$$e^{ikx} = coskx + isinkx \quad (5.57)$$

Then, f(x) can be expressed as

$$f(x) = \int_{-\infty}^{\infty} A_k e^{ikx} \quad (5.58)$$

where,

$$A_k = (1/2\pi)\int_{-\pi}^{\pi} f(x)\, e^{-ikx} dx \quad (5.59)$$

Fourier analysis of a numerical scheme: It was showed that any numerical scheme can be expressed as follows

$$C_j^{n+1} = gC_j^n \quad (5.60)$$

where g denotes the amplification factor. It was previously demonstrated that if $g \leq 1$ then any numerical scheme is stable, and vice versa. We can use the Fourier analysis approach to determine the stability condition of any numerical scheme. An example is shown below.

Explicit time/centered space: Let us assume that using Fourier series, C_j^n can be expressed as

$$C_j^n = Ae^{ik\Delta x} \tag{5.61}$$

where Δx is the spatial discretization. Therefore, C_{j+1}^n and C_{j-1}^n can be expressed as

$$C_{j+1}^n = Ae^{ik(j+1)\Delta x} \tag{5.62}$$

and

$$C_{j-1}^n = Ae^{ik(j-1)\Delta x} \tag{5.63}$$

Using Eq. 5.64, C_j^{n+1} can be expressed as in Equation (5.76)

$$C_j^{n+1} = gAe^{ik\Delta x} \tag{5.64}$$

Inserting Equation (5.66) and Equation (5.68) into Equation (5.49), we find

$$gAe^{ikj\Delta x} = Ae^{ikj\Delta x} - (\gamma/2)[Ae^{ik(j+1)\Delta x} - Ae^{ik(j-1)\Delta x}] \tag{5.65}$$

by dividing with $Ae^{ikj\Delta x}$, we find

$$g = 1 - (\gamma/2)(e^{ik\Delta x} - e^{-ik\Delta x}) \tag{5.66}$$

or,

$$g = 1 - \gamma\, i\sin k\Delta x \tag{5.67}$$

From the above equation, we can find the magnitude of the amplification factor g is

$$|g_a| = (1 + \gamma^2\sin^2 k\Delta x)^{0.5} \tag{5.68}$$

From Equation (5.72) we can see that for any value of γ, $|g_a| > 1$, and therefore, the centered difference method is unconditionally unstable when solving the pure advection equation.

5.9. Steps for Finite Difference Solutions

To set up, solve, and achieve a computational hydraulic solution for the flow problem, one must move across these interface fronts by carrying out a few simple tasks. The following are the steps in the solution procedure:

1. The pre-processing of the problem.
   a. Creating the geometry
   b. Generating the mesh
   c. Choosing the fluid and physical attributes
   d. Specifying the boundary conditions

2. A numerical solution
   a. Initialization and control of the solution
   b. Monitoring convergence

3. Post-processing result report and visualization.

5.10. Numerical Solutions of Differential Equations

5.10.1. Example of Finite Difference Approximation

We have a function given as $f(x) = 3x^3 + 2x^2$ for $x = 5$; find analytical and numerical solutions for the first and second derivatives.

Solution: Analytical solutions;

$$\frac{df}{dx} = 9x^2 + 4x = 225$$

$$\frac{d^2f}{dx^2} = 18x + 4 = 94$$

Finite difference solution: Choice $\Delta x = 0.1$. Use the formulas $x_{i+1} = x + \Delta x$, and $x_{i-1} = x - \Delta x$

For x =5, we can do the calculations as:

$x_{i+1} = 5.1$, $x_i = 5.0$, $x_{i-1} = 4.9$

Using forward finite differences;

$$\frac{df}{dx} = \frac{f(i+1) - f(i)}{\Delta x}$$
$$= \frac{3 \times 5.1 \times 5.1 \times 5.1 + 2 \times 5.1 \times 5.1 - 3 \times 5.0 \times 5.0 \times 5.0 - 2 \times 5.0 \times 5.0}{0.1}$$
$$= 249.73$$

Using backward finite differences;

$$\frac{df}{dx} = \frac{f(i) - f(i-1)}{\Delta x}$$
$$= \frac{3 \times 5.0 \times 5.0 \times 5.0 + 2 \times 5.0 \times 5.0 - 3 \times 4.9 \times 4.9 \times 4.9 - 2 \times 4.9 \times 4.9}{0.1}$$
$$= 240.33$$

Using centered finite differences;

$$\frac{df}{dx} = \frac{f(i+1) - f(i-1)}{2\Delta x}$$
$$= \frac{3 \times 5.1 \times 5.1 \times 5.1 + 2 \times 5.1 \times 5.1 - 3 \times 4.9 \times 4.9 \times 4.9 - 2 \times 4.9 \times 4.9}{0.2} = 245.03$$

$$\frac{d^2f}{dx^2} = \frac{f(i+1) - 2f(i) + f(i-1)}{\Delta x^2}$$
$$= \frac{(3 \times 5.1 \times 5.1 \times 5.1 + 2 \times 5.1 \times 5.1 - 2 \times 3 \times 5.0 \times 5.0 \times 5.0 - 2 \times 2 \times 5.0 \times 5.0 + 3 \times 4.9 \times 4.9 \times 4.9 + 2 \times 4.9 \times 4.9)}{0.01}$$
$$= 94$$

Chapter 6

Boundary and Initial Conditions

6.1. Introduction

In solving fluid flow equations that depend on space and time, both boundary and initial conditions should be known. If the flow is not time dependent, that is steady state condition, it is necessary to define the boundary conditions only. The variables in the equation that express the problem under these boundary and initial conditions must be determined at the beginning of the solution. The three main equations of fluid mechanics, conservation of mass, conservation of momentum and conservation of energy, are valid in hydraulics as well as in all other branches of physics. However, since the initial and boundary conditions of each problem are different, if the problem is solved using any of these equations and considering the specific boundary and initial conditions of the problem, the solution has a real physical meaning (Roache, 1979; Chapra and Canale, 1986; Golub and Ortega, 1993; Kumar, 2010).

Boundary conditions can be defined in the solid walls surrounding the flow as well as the boundary conditions at the entrance and exit of the system. In addition, in some problems to be solved, they may have some conditions such as the axis of symmetry or the surface of the free flow or the moving wall. At these boundaries, boundary conditions also should be defined. Briefly, in mathematical modeling, boundary conditions are required to define at every boundary surrounding the flow domain (Jenkins et al., 1983).

Boundary conditions can be classified as: 1. Neumann boundary condition, 2. Dirichlet boundary condition, and 3. Robin or mixed boundary conditions. If we prescribe the value of a function, f, at the boundary, this boundary condition is called as Dirichlet boundary condition.

When we specify the normal derivative of the function, as $\partial f/\partial n$, on a surface, this boundary condition is called as Neumann boundary condition.

In mixed boundary conditions, some f values and some gradient values are known around the solution domain. Mixed boundary is the combination of Dirichlet and Neumann conditions essentially weighted combination of Dirichlet and Neumann.

In computational hydraulics, we can examine some different types of boundary conditions which can be used for different variables as seen on Table 6.1. In this table, u, v, and w are velocity components, ψ is the stream function, ξ is the vorticity, h is the water depth, Q is the discharge, C is the concentration (Franke et al., 1987).

Table 6.1. General boundary values need

Unknown variables	Example of equation	Boundary values needed for
Velocity components, and water depth	Primitive variables formulations, Navier-Stokes and St. Venant equations	u, v, w, h
Stream function	Laplace Equation	Ψ
Stream function and vorticity	Poisson Equation	ψ, ξ
Water discharge	Continuity Equation	Q, A
Concentration and velocity	Dispersion Equation	c, u, w

6.2. Governing Equations

In this section, we want to take into account various formulation types.

a. Formulas for Primitive Variables: We look at the fundamental equations used in fluid dynamics and hydraulics in Chapter 2. Here, we look at a few equations while taking various primitive variable types into account. For the flow of an incompressible fluid in two dimensions, we can formulate simple equations. The continuity equation and two momentum equations are listed below.

Continuity Equation:

$$\frac{\partial u}{\partial x} + \frac{\partial v}{\partial y} = 0 \tag{6.1}$$

Momentum Equation:

$$\frac{\partial u}{\partial t} + u\frac{\partial u}{\partial x} + v\frac{\partial u}{\partial y} = -\frac{\partial p}{\partial x} + \frac{1}{R_e}\left(\frac{\partial^2 u}{\partial x^2} + \frac{\partial^2 u}{\partial y^2}\right) \tag{6.2}$$

$$\frac{\partial v}{\partial t} + u\frac{\partial v}{\partial x} + v\frac{\partial v}{\partial y} = -\frac{\partial p}{\partial y} + \frac{1}{Re}\left(\frac{\partial^2 v}{\partial x^2} + \frac{\partial^2 v}{\partial y^2}\right) \quad (6.3)$$

where u, and v are velocity components, p is the pressure, R_e is the Reynolds number.

These are the Navier-Stokes equations in two dimensional flow.

b. Saint-Venant Equations:

$$\frac{\partial u}{\partial t} + h\frac{\partial u}{\partial x} + u\frac{\partial h}{\partial x} = 0 \quad (6.4)$$

$$\frac{1}{g}\frac{\partial u}{\partial t} + \frac{u}{g}\frac{\partial u}{\partial x} + \frac{\partial h}{\partial x} + S_f - S_0 = 0 \quad (6.5)$$

where u, and v are velocity components, h is the water depth, g is the gravitational acceleration, S_f is the friction slope, and S_o is the bottom slope.

c. Laplace Equation

$$\frac{\partial^2 \psi}{\partial x^2} + \frac{\partial^2 \psi}{\partial y^2} = 0 \quad (6.6)$$

where ψ is the stream function.

d. Poisson Equation:

$$\frac{\partial^2 \psi}{\partial x^2} + \frac{\partial^2 \psi}{\partial y^2} = \xi \quad (6.7)$$

where, ψ is the stream function, ξ is the vorticity.

e. Continuity Equation:

$$\frac{\partial A}{\partial t} + \frac{\partial Q}{\partial x} = 0 \quad (6.8)$$

where A is the cross-sectional area, Q is the discharge.

f. Dispersion Equation:

$$u\frac{\partial c}{\partial x} + w\frac{\partial c}{\partial y} = \varepsilon\frac{\partial^2 c}{\partial y^2} \tag{6.9}$$

where, u and w are the velocity components, c is the concentration.

6.3. Definition of Stream Function and Velocity Potential Function

Some streamlines of a flow are shown in Figure 6.1. In this figure, u is the horizontal velocity component of the flow, v is the vertical velocity component of the flow, and ψ is the stream function.

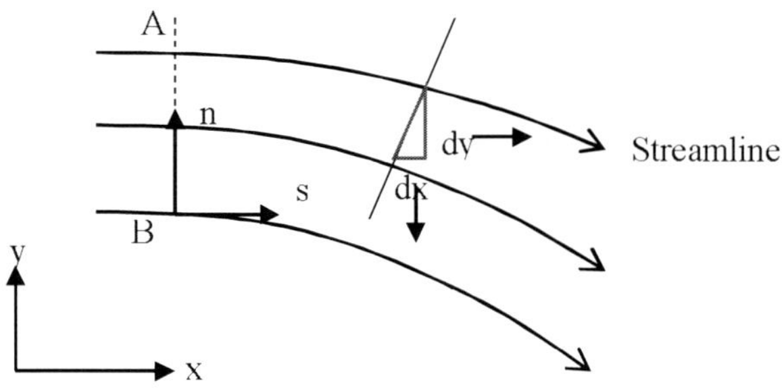

Figure 6.1. Definition of streamlines.

The fluid rate between two stream lines separated by an elemental distance is

$$d\psi = udy - vdx \tag{6.10}$$

where $d\psi$ is an exact differential and can be written:

$$d\psi = \frac{\partial\psi}{\partial x}dx + \frac{\partial\psi}{\partial y}dy \tag{6.11}$$

Thus, velocity components are,

$$u = \frac{\partial\psi}{\partial y} \tag{6.12}$$

$$v = -\frac{\partial\psi}{\partial x} \tag{6.13}$$

Velocity components can be written in finite difference expression,

$$u_{i,j} = \frac{\psi_{i,j+1} - \psi_{i,j-1}}{2\Delta y} \tag{6.14}$$

$$v_{i,j} = -\frac{\psi_{i+1,j} - \psi_{i-1,j}}{2\Delta x} \tag{6.15}$$

The flow rate between points A and B in Figure 6.1 can be written as:

$$Q = Dq = D\int_A^B v_n dn = D\int_A^B udy - vdx = D\int_A^B \frac{\partial\psi}{\partial x}dx + \frac{\partial\psi}{\partial y}dy = D\int_A^B d\psi \tag{6.16}$$

$$Q/D = q = \psi_A - \psi_B \tag{6.17}$$

where q is the flow rate per unit wide,

D is the third dimension of the flow domain which is perpendicular to this paper,

v_n is the velocity component normal to the flow direction.

Equation (6.17) means that the difference value of stream functions between points A and B is equal to flow rate per unit width through these points.

6.4. Boundary Conditions for the Primitive Equations

As shown in Figure 6.2, we want to discuss the boundary conditions for a primitive equation employing an axially symmetric flow domain for flow over a reservoir spillway.

In primitive equation we have to describe the boundary conditions in velocity components. In an axial flow as shown in Figure 6.2, there are different types of boundary conditions as wall boundaries, symmetry boundary, upstream and downstream boundaries, and upper boundary (Notdström, 2015).

Wall boundaries: Wall boundary condition is called also solid wall condition. Solid wall bounds the flow domain as a rigid body. On solid wall, the tangential velocity component is equal to wall velocity, and the normal velocity component is zero.

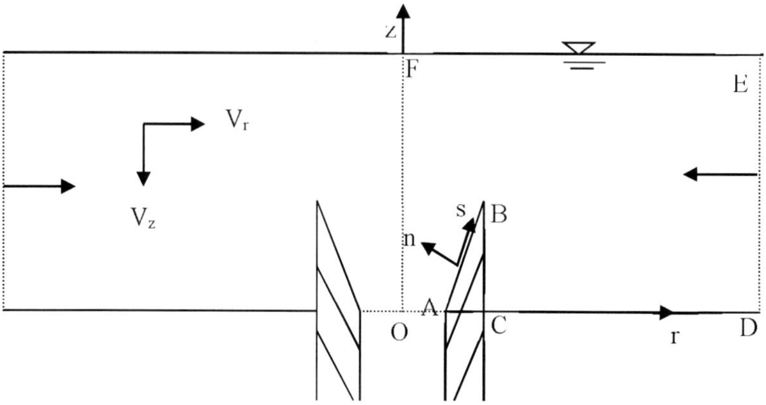

Figure 6.2. An axial symmetric flow domain: Flow over a shaft spillway.

Symmetry boundaries: On symmetry boundary, the velocity components can be defined as: 1. Normal velocity to symmetry plane is zero; 2. Normal gradient of all variables to symmetry plane is zero.

Upstream and Downstream boundary conditions: These sometimes are called inlet and outlet boundaries. Boundary conditions permits flow to enter and exit solution domain, generally. Some of them are examined in this chapter.

Upstream boundary: If velocity properties are known at the inlet, these velocity values are used as boundary conditions. If not known, a velocity profile is estimated at inlet boundary. Uniform, linear or other velocity distributions can be selected for the inlet velocity. In these cases, the inlet boundary of the domain should not be chosen close. Otherwise, precision cannot be obtained in computations.

Outflow boundary: For flow exit, outflow boundary conditions are also necessary. For outflow boundary condition, the similar statements can be said as upstream boundary condition in modelling.

For an axially symmetric flow, boundary condition for primitive equation can be descried as following.

(a) Solid walls (Wall boundaries): Boundaries by the no slip condition can be written on AB,

$$v_n = 0 \tag{6.18}$$

$$v_s = 0 \tag{6.19}$$

where v_n is the velocity component normal to the wall, v_s is the velocity component tangential to the wall. On BC:

$$v_r = 0 \tag{6.20}$$

$$v_z = 0 \tag{6.21}$$

where v_r is the velocity component in r direction, v_z is the velocity component in z direction.

On CD: as the same as on BC, i.e., $v_r = 0$, and $v_z = 0$.

(b) Symmetry boundaries: on OF, v_z = symmetric and, $v_r = 0$.
(c) Upper boundary: Upper boundary can be written as, on EF, $v_z = 0$.
(d) Upstream Boundary: We can select any velocity distribution, on ED, $v_z = 0$.

As shown in Figure 6.3, we can define the following for various velocity distributions:

a. Uniform velocity distribution on ED: $v_z = 0$, and v_r = constant,
b. Linear velocity distribution on ED: $v_z = 0$, and $v_r = az + b$, or,
c. Parabolic velocity distribution on ED: $v_z = 0$, and $v_r = az^2 + bz + c$.

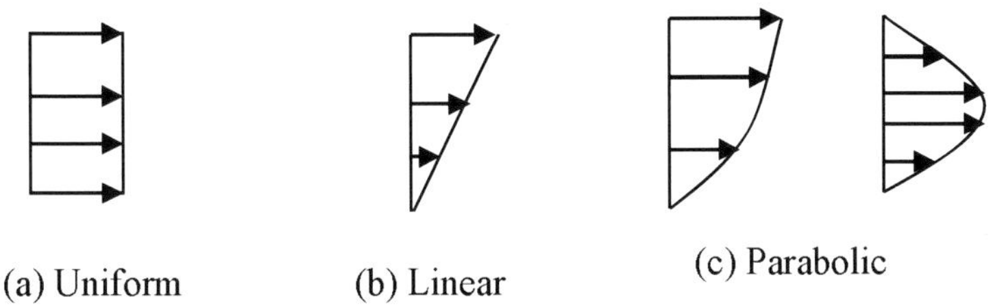

Figure 6.3. Different velocity distributions for upstream boundary.

(e) Outflow boundary: on OA, $v_r = 0$.

As shown in Figure 6.4, there are various options for the outflow boundary, including:

a. Uniform velocity distribution on OA, $v_r = 0$, and v_z = constant,
b. Linear velocity distribution on OA, $v_r = 0$, and $v_z = az + b$, or,
c. Parabolic velocity distribution on OA. $v_r = 0$, and $v_z = az^2 + bz + c$.

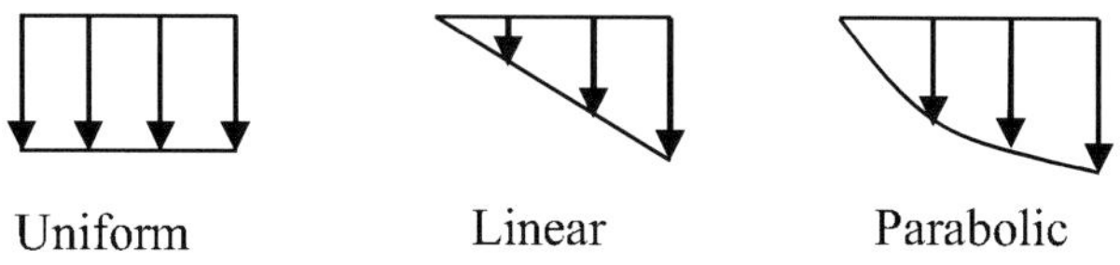

Figure 6.4. Different velocity distributions for outflow boundary.

6.5. Determination of Velocity Components from Nondimensional Stream Function

In computational hydraulics, we obtained stream function values in some cases. Then, we want to calculate the velocity components from stream function values. For instance, we consider a horizontal sharp corner elbow flow as can be seen in Figure 6.5.

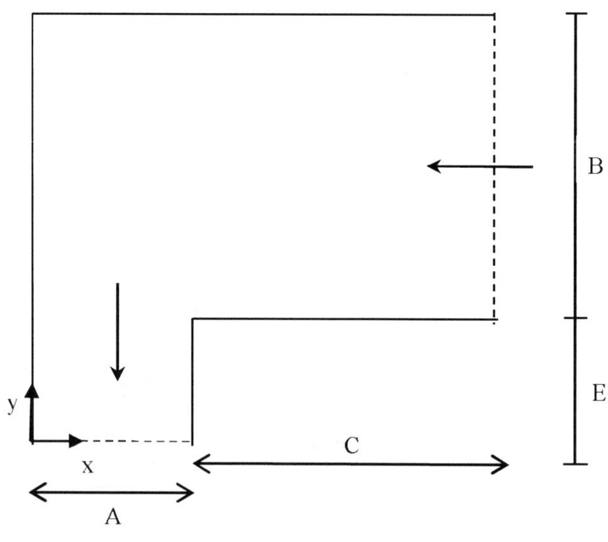

Figure 6.5. Horizontal sharp corner elbow flow scheme.

It can be written a relationship between velocity components and steam function such as Equation (6.12) and Equation (6.13), where u and v are velocity components and ψ is the stream function.

In order to use non-dimensional stream function values, we want to define non-dimensional form of equation by using the following definitions applying the characteristic length as B:

$$\frac{y}{B} = y_1 \tag{6.22}$$

$$\frac{x}{B} = x_1 \tag{6.23}$$

$$\frac{\psi}{\psi_{max}} = F \tag{6.24}$$

where ψ_{max} is the maximum value of stream function, F is the dimensionless stream function,

Using Equations (6.22), (6.23) and (6.24) we can write,

$$\partial y = B\partial y_1 \tag{6.25}$$

$$\partial x = B\partial x_1 \tag{6.26}$$

$$\partial\psi = \psi_{max}\,\partial F \tag{6.27}$$

Therefore, from Equations (6.12) and (6.13), one can write:

$$u = \frac{\psi_{max}}{B}\frac{\partial F}{\partial y} \tag{6.28}$$

$$v = -\frac{\psi_{max}}{B}\frac{\partial F}{\partial x} \tag{6.29}$$

On the other hand, from Equation (6.17), we can write,

$$\Delta Q = D\Delta\psi \tag{6.30}$$

or

$$\Delta Q/D = \Delta q = \Delta\psi \tag{6.31}$$

In a similar way, we can write,

$$Q_{max} - Q_0 = D(\psi_{max} - \psi_0) \tag{6.32}$$

When $Q_0 = 0$ and $\psi_0 = 0$, then $Q_{max} = Q$, one can write,

$$Q = D\psi_{max} \tag{6.33}$$

and

$$q = \psi_{max} \tag{6.34}$$

This equation valid for Cartesian coordinates. For an axial flow in cylindrical coordinates we can write,

$$dQ = 2\pi r d\psi \tag{6.35}$$

$$Q = 2\pi r \Delta\psi \tag{6.36}$$

or

$$\Delta q = 2\pi\psi_{max} \tag{6.37}$$

Therefore, the discharge per unit width may be determined by taking the difference of any two streamlines values.

We define a constant k as,

$$\frac{\psi_{max}}{B} = \frac{Q}{DB} = k \tag{6.38}$$

We can obtain as

$$u = k\frac{\partial F}{\partial y_1} \tag{6.39}$$

or in finite difference expression,

$$u = k\frac{F_{i,j+1} - F_{i,j-1}}{2\Delta y_1} \tag{6.40}$$

Similarly, we can write,

$$v = -k\frac{\partial F}{\partial x_1} \tag{6.41}$$

or in finite difference expression,

$$v = -k\frac{F_{i+1,j} - F_{i-1,j}}{2\Delta x_1} \tag{6.42}$$

6.6. Boundary Conditions for the Stream Function Equations

Boundary conditions for the stream function equation will be discussed by an axial symmetric flow example as shown in Figure 6.6. This figure describes an axial symmetric flow over a shaft spillway in a reservoir. In this problem, potential flow equation, as known Laplace Equation, can be written as

$$\frac{\partial^2\psi}{\partial r^2} + \frac{1}{r}\frac{\partial\psi}{\partial r} + \frac{\partial^2\psi}{\partial z^2} = 0 \tag{6.43}$$

where ψ is the stream function.

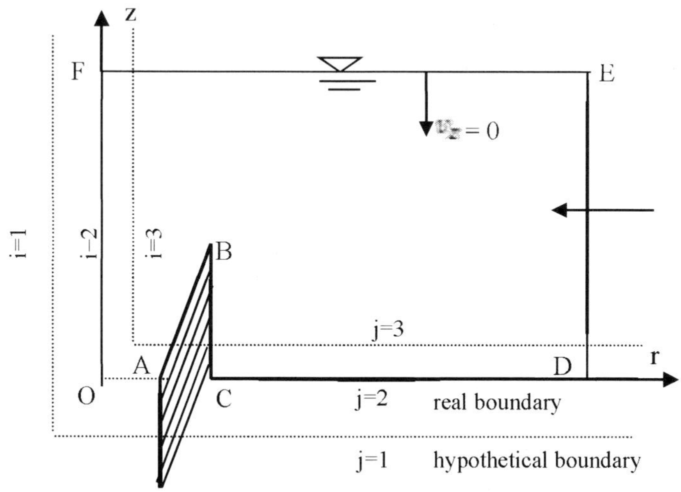

Figure 6.6. Scheme of half part of the flow domain as in Figure 6.2.

In an axial symmetric flow, velocity components, v_r and v_z, can be written as

$$v_r = \frac{1}{r}\frac{\partial\psi}{\partial z} \qquad (6.44)$$

$$v_z = -\frac{1}{r}\frac{\partial\psi}{\partial r} \qquad (6.45)$$

Boundary conditions on AB, BC, and CD: As no-slip condition, velocity components are zero on wall boundaries, $v_r = 0$, $v_z = 0$, therefore,

$$\Psi = \text{Constant} \qquad (6.46)$$

In non- dimensional form, we define the dimensionless stream function as $F= \psi/\psi_{max}$. This boundary is the lower part of the flow. Therefore

$$F = 0 \qquad (6.47)$$

Boundary conditions on EF and FO: This line is a streamline. So,

$$\frac{\partial F}{\partial r} = 0 \qquad (6.48)$$

Boundary condition on EF: $v_z = 0 = -(1/r)(\partial F/\partial r) = 0$, so,

$$F = \text{Constant} \qquad (6.49)$$

Boundary condition on FO: $v_r = 0 = -(1/r)(\partial F/\partial z) = 0$, so, F = Constant. This boundary (FO) is upper side of the flow domain, So,

$$F_{i,j} = 1 \qquad (6.50)$$

Boundary condition on DE: We assume a uniform velocity distribution on this boundary,

So, we can write as seen from Figure 6.7. v_z = Uniform, and v_r = Constant, $v_r = (1/r)(\partial F/\partial z)$ = Constant,

$$F = az + b \qquad (6.51)$$

This is a linear relationship. It can be written in finite difference form,

$$F_{i,j} = 2F_{i,j+1} - F_{i,j+2} \tag{6.52}$$

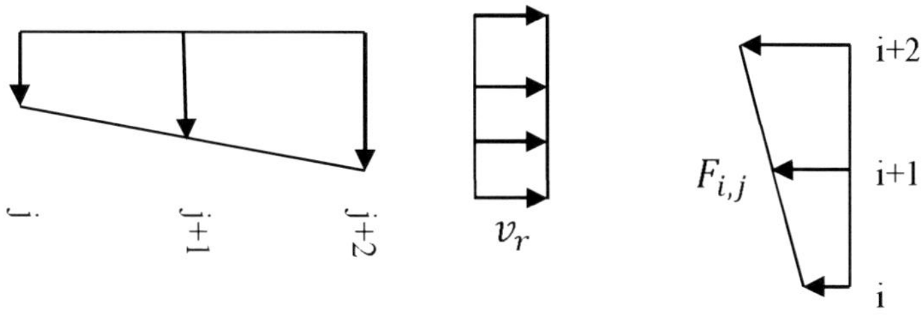

Figure 6.7. Velocity and stream function distribution on upstream boundary, DE.

Boundary condition on OA: This is the outflow boundary of flow domain. If we assume uniform velocity boundary, we can write v_z = uniform, $v_z = -(1/r)(\partial F/\partial r)$ = Constant. In finite difference form, one can write,

$$F_{i,j} = 2F_{i+1,j} - F_{i+2,j} \tag{6.53}$$

6.7. Boundary Conditions for the Vorticity and Stream Function Equations

6.7.1. General

It is easy to see some kind of plausible boundary conditions for ψ and ξ, but attempts to determine realistic, accurate, and stable methods can be highly frustrating. It has been found that the adequacy of any boundary condition, as determined by computational experiments, can depend on the Reynolds number, the interior-point differencing methods, the other boundary conditions, and sometimes on initial conditions.

No mathematically rigorous solutions are available for many of the problems. We will draw mainly upon perception intuition, and computational experimentation. Most of the computational experiments on boundary conditions have been carried out using simple two-level explicit methods for the vorticity transport equations. We will note the few known cases where the same boundary conditions caused instability when used with other methods.

Most of the boundary conditions given are either Dirichlet conditions or the Neumann conditions. No fluid dynamics computations have been performed to date with mixed type, where a weighted combination of function value and normal gradient are specified.

For description of boundary conditions, we consider the Poisson equation, Equation (6.7).

Boundary conditions will be discussed in relation to the planar 2-D backstep problem shown in Figure 6.8, which affords examples of all types of boundaries encountered in a block-rectangular region.

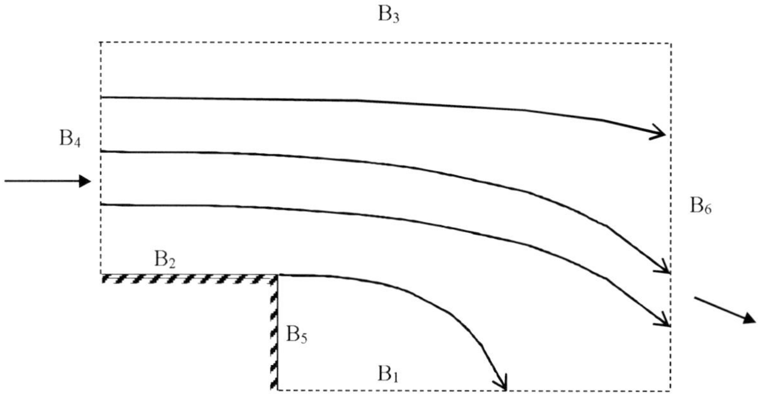

Figure 6.8. Boundary labeling.

Examples of all types of boundaries encountered in a rectangular region are as follows,

1. Walls (solid walls), B_2, B_5.
2. Symmetry boundaries, B_1.
3. Upper boundary (on free surface), B_3.
4. Upstream boundary, B_4.
5. Outflow boundary (or downstream boundary), B_6.

6.7.2. Walls in the First Mesh System

Solids walls appear in Figure 6.8 at boundary B_2, and at B_5. In the first mesh system, node values of ψ and ξ lie along the walls. Since the line of B_2-B_5-B_1

is a streamline, any constant value of ψ may be selected; the conventional choice is $\psi = 0$.

The wall vorticity is an extremely important evaluation. At no slip-boundaries, however, ξ is produced. The vorticity is obtained from the no-slip conditions. Using boundary B_2, we expand to $\psi_{i,jc+1}$ by a Taylor series out from the wall values, $\psi_{i,jc}$.

$$\psi_{i,jc+1} = \psi_{i,jc} + \left|\frac{\partial\psi}{\partial y}\right|_{i,jc} \Delta y + \frac{1}{2}\left|\frac{\partial^2\psi}{\partial y^2}\right|_{i,jc} \Delta y^2 + \frac{1}{6}\left|\frac{\partial^3\psi}{\partial y^3}\right|_{i,jc} \Delta y^3 + 0(\Delta y^4) \tag{6.54}$$

At this boundary, $\left|\frac{\partial\psi}{\partial y}\right|_{i,jc} = u_{i,jc} = 0$. By the no-slip condition, and $\left|\frac{\partial^2\psi}{\partial y^2}\right|_{i,jc} = \left|\frac{\partial u}{\partial y}\right|_{i,jc}$. Now, the definition of vorticity is

$$\xi = \partial u/\partial y - \partial v/\partial x \tag{6.55}$$

Along the wall, because v = constant (i.e., v = 0) along x. Thus, Substituting this into Equation (6.54) and solving for $\xi_{i,jc}$ with $\psi_{i,jc} = 0$ gives

$$\xi_{i,jc} = \frac{2\psi_{i,jc+1}}{\Delta y^2} + 0(\Delta y) \tag{6.56}$$

No matter what's boundary value or wall orientation is, we can write

$$\xi_w = \frac{2(\psi_{w+1} - \psi_w)}{\Delta n^2} + 0(\Delta n) \tag{6.57}$$

where Δn is the distance from (w+1) to (w) , normal to the wall.

This first-order form is used extensively. It is the safest form to use, and it often gives results essentially equal to higher order forms. It is suggested the following second-order form, obtainable from equation (6.55) by retaining the Δy^3 term. By differentiating the definition of vorticity, we obtain

$$\frac{\partial\xi}{\partial y} = \frac{\partial^2 u}{\partial y^2} - \frac{\partial^2 v}{\partial y\,\partial x} = \frac{\partial^3\psi}{\partial y^3} - \frac{\partial}{\partial x}\left(\frac{\partial v}{\partial y}\right) \tag{6.58}$$

From the continuity equation (6.1), $\partial v/\partial y = -(\partial u/\partial x)$. Substituting this above and evaluating at the wall, we obtain,

$$\left|\frac{\partial \xi}{\partial y}\right|_w = \left|\frac{\partial^3 \psi}{\partial y^3}\right|_w + \left|\frac{\partial^2 u}{\partial x^2}\right|_w \tag{6.59}$$

The second term $|\partial^2 u/\partial x^2|_w$, is zero by the no-slip condition (u = constant = 0). The term $|\partial \xi/\partial y|_w$ is evaluated by a first-order backward difference from (w+1).

Substituting this equation for $|\partial^3 \psi/\partial y^3|_w$ into equation (6.55) and solving for ξ_w gives Woods' method for all vorticity.

$$\xi_w = \frac{3(\psi_{w+1} - \psi_w)}{\Delta n^2} - \frac{1}{2}\xi_{w+1} + 0(\Delta n^2) \tag{6.60}$$

Second-order formulations for ξ_w in axisymmetric coordinates were evaluated In terms of the local radius of curvature r_c, the expression was

$$\xi_w = \frac{\frac{3\psi_{w+1}}{\Delta n^2} - \frac{1}{2}\xi_{w+1} + \Delta n^2 Re \frac{\partial \xi_w}{\partial t}/24}{1 - \frac{\Delta n}{2r_c} + \frac{3\Delta n^2}{8r_c^2}} + 0(\Delta n^2) \tag{6.61}$$

For $r_c \rightarrow \infty$, the equation appropriate for a flat plate is obtained, and is seen to be just expression (6.60) plus the time-dependent term.

A more frequently used second-order form, with the condition of $|\partial \psi/\partial y|_w = u_w = 0$. The result is

$$\xi_w = \frac{-7\psi_w + 8\psi_{w+1} - \psi_{w+2}}{2\Delta n^2} + 0(\Delta n^2) \tag{6.62}$$

Another equation based on a fourth-order Lagrange interpolating polynomial is derived as,

$$\xi_w = \frac{-85\psi_w + 108\psi_{w+1} - 274\psi_{w+2} + 4\psi_{w+3}}{18\Delta n^2} + 0(\Delta n^2) \tag{6.63}$$

A consistent formulation requires that the velocities at two points near the wall be specially evaluated as

$$u_{w+2} = \frac{14\psi_1 - 36\psi_2 + 18\psi_3 + 4\psi_4}{18\Delta y^2} \tag{6.64}$$

and furthermore that the term $\partial^2\psi/\partial y^2$ in the Poisson equation must be evaluated as

$$\left|\frac{\partial^2\psi}{\partial y^2}\right|_{w+1} = \frac{29\psi_w - 54\psi_{w+1} + 27\psi_{w+2} - 2\psi_{w+3}}{18\Delta y^2} \tag{6.65}$$

$$\left|\frac{\partial^2\psi}{\partial y^2}\right|_{w+2} = \frac{11\psi_w - 27\psi_{w+2} + 16\psi_{w+3}}{18\Delta y^2} \tag{6.66}$$

6.7.3. Sloping Wall Boundary

For the case of sloping wall lying diagonally along the mesh, as shown in Figure 6.9, the wall vorticity, may be evaluated from the first-order form Equation (6.66) as follows.

$$\xi_{wm} = \frac{2(\psi_m - \psi_w)}{\Delta n^2} \tag{6.67}$$

$$\Delta n^2 = [\Delta y \cos\theta_w]^2 = \frac{\Delta x^2}{1+\beta^2} \tag{6.68}$$

where $\beta = \Delta x/\Delta y$ is the mesh aspect ratio. Then, the wall value, at node w_l is determined by interpolation along the wall.

The method is stable and apparently accurate. For the particular case of a 45° wall inclination ($\beta = 1$) this procedure gives, for a point (i_w, j_w) as shown in Figure 6.9,

$$\xi_{iw,jw} = \frac{2(\psi_{iw-1,jw} + \psi_{iw,jw+1} - 2\psi_{iw,jw})}{\Delta^2} \tag{6.69}$$

where $\Delta = \Delta x$. This method appears to be more accurate than using

$$\xi_{iw,jw} = \frac{(\psi_{iw-1,jw+1} - \psi_{iw,jw})}{\Delta^2} \tag{6.70}$$

Since the ψ values used in Equation (6.69) are twice as far from the wall as those in Equation (6.69). It may also be useful to model a "slip" wall condition with the viscous equations. This implies that the boundary layer is less than Δy thick. This condition may be heuristically simulated by using a "slip" wall condition. The ψ value is specified, and the vorticity is evaluated from a Neumann condition,

$$\xi_w = \xi_{w+1} \tag{6.71}$$

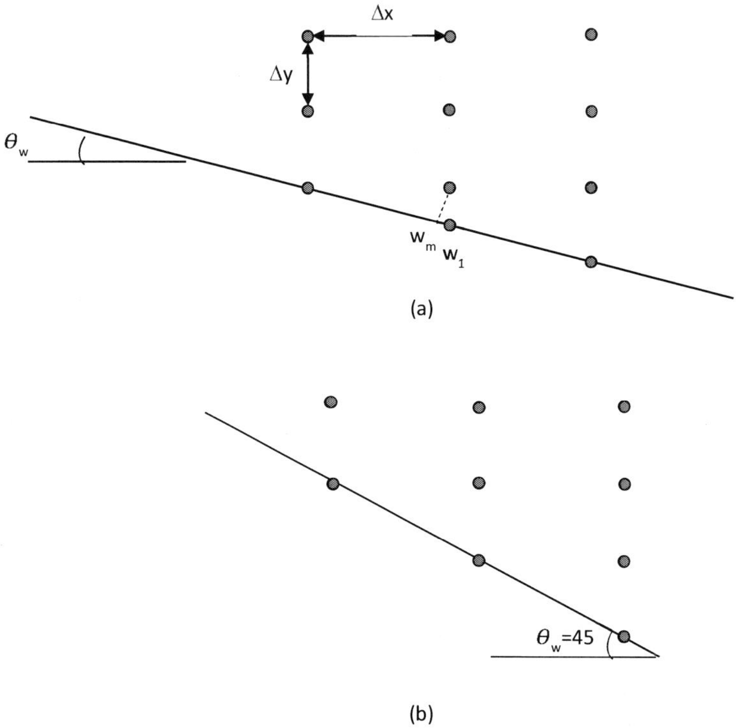

Figure 6.9. Evaluation of wall vorticity on a sloping wall.

6.7.4. Symmetry Boundaries

If the centerline boundary B_1 in Figure 6.8 were a splitter plate, it would have the same boundary conditions as a solid no-slip wall. Even if the figure represents a symmetric half of a base flow problem, $\psi = 0$ is still required, implying that only subcritical wake solutions are relevant. In this case, B_1 represents a slip splitter plate, and the boundary condition on vorticity is very simple. Because $v = 0$ everywhere on the centerline, $\partial v/\partial x = 0$ and u is symmetric above and below the centerline, $\partial u/\partial y = 0$. Thus,

$$\xi = 0 \tag{6.72}$$

We cannot use symmetry conditions on ψ derivatives in addition to the condition $\psi = 0$, this would over specify the Poisson solution.

A symmetry of $\xi = 0$ was used along a 45^0 line between a wall and the centerline of an impinging normal jet. Symmetry in this case applies strictly only to potential flow equation, but was used in here as a High-Reynold approximation.

6.7.5. Upper Boundary

The upper boundary, B_3 in Figure 6.8, can pose an interesting problem. Of course, we can choose to model a physical problem whose upper boundary condition is obvious; for example, if we choose to model an unsymmetrical channel expansion, then B_3 is a solid no-slip wall and those equations for the vorticity apply. The value ψ (B_3) is constant and may be evaluated by integrating the input u-velocity profile at B_4.

A considerable improvement accrues if a moving wind tunnel wall is modeled. They specified $v = 0$, and $u = U_0$, where U_0 may be interpreted as a "free stream" velocity. The boundary conditions in the system would then be those for a no-slip moving wall ψ = constant, and ξ evaluated from

$$\xi_{ij} = \frac{2(\psi_{i,j-1} - \psi_{ij} - U_0\Delta y)}{\Delta y^2} + 0(\Delta y) \tag{6.73}$$

or

$$\xi_{ij} = \frac{3(\psi_{i,j-1} - \psi_{ij} - U_0\Delta y)}{\Delta y^2} - \frac{1}{2}\xi_{i,j-1} + 0(\Delta y^2) \tag{6.74}$$

Another improvement over modeling a physical wind tunnel wall is to model a frictionless wind tunnel wall. The value of $\psi(B3)$ is still constant and vorticity is evaluated using the concept of "image" system gives ξ (B_3) = 0. The boundary can be specified the far-field condition of $\xi = 0$ with $\partial\psi/\partial y \equiv u = U_0$, and $\partial\psi/\partial y \equiv v = U_0$. The U_0 condition was set by writing the Neumann condition along B_3 as,

$$\psi_{i,J} = \psi_{i,J-1} + U_0\Delta y \tag{6.75}$$

If the boundary line B_3 is taken j = J, Equation (6.83) is only first-order accurate. But by simply taking B_3 to lie between j = J and j = J-1, which we donate by (i, J-1/2), Equation (6.83) becomes "second-order" accurate. The condition with Equation (6.83) just gives at B3,

$$\psi_{i,J} = \psi_{1,J} \quad (6.76)$$

or, if B_3 is at (i,J-1/2), it gives

$$\psi_{i,J} = \psi_{i,J-1} + U_0 \Delta y \quad (6.77)$$

In these equations, $\psi_{i,J}$ or $\psi_{i,J-1}$ are values at the top of the inflow boundary, B_4 in Figure 6.8.

We can set another boundary condition at B_3 as,

$$\xi_{i,j} = \xi_{i,j-1} \quad (6.78)$$

It was computed the boundary-layer separation bubble on a flat plate by specifying the u-velocity at B3 as,

$$u = U_0(x) = a + bx \quad (6.79)$$

up to some x_1 such that $x < x_1 < -a/b$, and u = constant. Along with the condition $\xi = 0$ through B_3, this specification causes the boundary layer along B_1 to separate and to subsequently reattach.

6.7.6. Upstream Boundary

The upstream boundary B_4 in Figure 6.8 cannot have a unique solution because its characteristics will change depending on the physical flow upstream of the mesh and upon completely specified the inflow boundary conditions.

It was used a less restrictive input for the mesh boundary well ahead of the cylinder. It is required $v(1, j) = 0$, which gives condition

$$\psi_{1,j} = \psi_{2,j} \quad (6.80)$$

This allows u(1, J) to develop as a part of the computation which their upper boundary treatment, $u = U_0$ is specified at B_3, but ψ (B_3) also develops.

The most successful method found was to approximate

$$\frac{\partial}{\partial x}\left(\frac{\partial v}{\partial x}\right)_{1,j} = 0 \tag{6.81}$$

by

$$\frac{\partial v}{\partial x}\bigg|_{i,j} = \frac{\partial v}{\partial x}\bigg|_{2,j} = \frac{\partial}{\partial x}\left(-\frac{\partial \psi}{\partial x}\right)_{2,j} = -\frac{\partial^2 \psi}{\partial x^2}\bigg|_{2,j} \tag{6.82}$$

or finally

$$\frac{\partial v}{\partial x}\bigg|_{1,j} = -\frac{\psi_{1,j} + \psi_{3,j} - 2\psi_{2,j}}{\Delta x^2} \tag{6.83}$$

Thus, the input profile fixes ψ and $\partial u/\partial y$ while $\xi = \partial u/\partial y - \partial v/\partial x$ is set from the input $\partial u/\partial x$ and Equation (6.80).

The alternative method is to discretize the u(B_3) solution and the evaluate $\psi_{i,j}$ in a manner compatible with the velocity differencing, as in

$$u_{1,j} = \frac{\psi_{1,j+1} - \psi_{1,j-1}}{2\Delta y} \tag{6.84}$$

This method would give

$$\psi_{i,jc} = 0 \tag{6.85}$$

$$\psi_{i,jc+1} = \frac{1}{2} u_{1,jc+1} \Delta y \tag{6.86}$$

$$\psi_{1,j} = u_{1,j-1} 2\Delta y + \psi_{1j-2} \tag{6.87}$$

This second method gives the correct velocities by equation (6.71), but results in a discretization error for $\psi_{1,j}$ of course, both methods converge for $\Delta y \to 0$.

6.7.7. Outflow Boundary

The evaluation of ψ and ξ at the outflow boundary B_6 of Figure 6.8 is one of the most interesting computational boundary problems. We must somehow neglect the details of farther downstream flow and obtain realistic answers upstream.

In a two-dimensional channel flow, it is specified that $v = \partial\psi/\partial x = 0$ and $\partial\xi/\partial x = 0$ at outflow boundary. It was found that this less restrictive condition to be an improvement over the parabolic flow specification, as expected, allowing a considerably shorter computational mesh for comparable accuracy in the region of interest. With the present notation of max i= I, and dropping the j-index their conditions are

$$\psi_I = \psi_{I-1} \tag{6.88}$$

$$\xi_I = \xi_{I-2} \tag{6.89}$$

More restrictive conditions of

$$\psi_I = \psi_{I-1} \tag{6.90}$$

$$\xi_I = 0 \tag{6.91}$$

It was also developed even less restrictive outflow boundary conditions. They approximated $\partial\xi/\partial x = 0$ and $\partial v/\partial x = 0$. Since $v = -\partial\psi/\partial x$, this second condition implies $\partial^2\psi/\partial x^2$ which was approximated by linear extrapolation out to i= I. For

$$\xi_I = \xi_{I-1} \tag{6.92}$$

$$\xi_I = \xi_{I-1} \tag{6.93}$$

$$\psi_I = 2\psi_{I-1} - \psi_{I-2} \tag{6.94}$$

To second-order accuracy, these really set $\partial\xi/\partial x = 0$ at (I-1/2) and $\partial^2\psi/\partial x^2 = 0$ at (I-1). In some studies, the following expressions at outflow were used:

$$\psi_I = \psi_{I-4} - 2\psi_{I-3} + 2\psi_{I-1} \tag{6.95}$$

$$\xi_I = \xi_{I-4} - 2\xi_{I-3} + 2\xi_{I-1} \tag{6.96}$$

6.8. Irregular Boundaries

A general problem, which arises due to the morphological complexity of the solution domains for all types of parabolic, hyperbolic and elliptic equations, is the approximation of curved boundaries and the adaptation of the difference equations on those boundaries. This is a very frequent situation in hydraulics where the geophysical flow domains have a complex geometry.

The termination, for example, of the flow field at point 1 of Figure 6.10, instead of point 0 of the grid, requires either the artificial extension of the field to 0 or the proper adaptation of the finite differences formulae as described below.

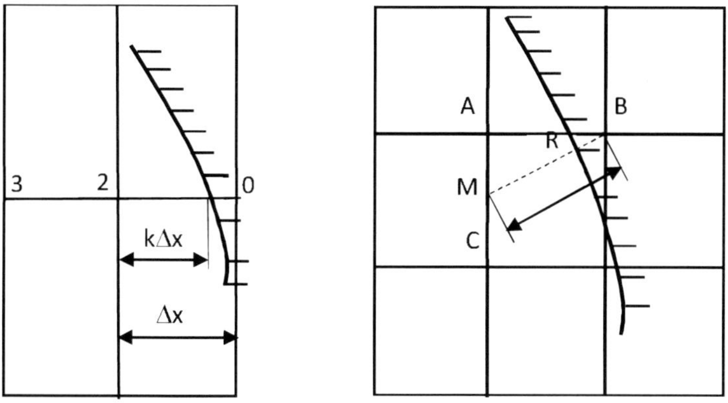

Figure 6.10. Irregular Boundaries.

Through the expansion in Taylor Series of the field variable at points 1 and 3 we can approximate the derivatives at 2 as follows:

$$\left.\frac{\partial f}{\partial x}\right|_2 = \frac{1}{\Delta X}\left(\frac{f_1}{k+1} - \frac{kf_3}{1+k} - \frac{(1-k)f_2}{k}\right) \tag{6.97}$$

$$\left.\frac{\partial^2 f}{\partial x^2}\right|_2 = \frac{1}{\Delta x^2}\left(\frac{2f_1}{k(k+1)} + \frac{2f_3}{1+k} - \frac{2f_2}{k}\right) \tag{6.98}$$

The f_2 value is expressed through Equations (6.97) and (6.98) when the f_1 value is given. When the boundary condition refers to the gradient of f

normal to the boundary, then, and according to the notation of Figure 6.10, a fictitious point B is introduced in the flow domain and the $\partial f/\partial x$ derivative at point R is approximated by the difference

$$\left.\frac{\partial f}{\partial x}\right|_R = \frac{f_B - f_M}{d_{BM}} \tag{6.99}$$

The f_M value is computed by means of interpolation between the f_A and f_C values. Equation (6.99) is used for the computation of the unknown f_R value.

6.9. Boundary Conditions for Sharp Corners

In solution, we need some boundary conditions for sharp corners of the flow domain as seen in Figure 6.11. The labeling of the grid points are defined as shown in Figure 6.12.

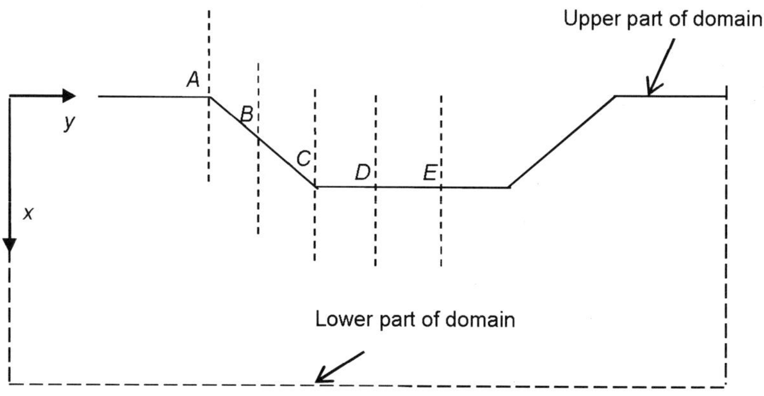

Figure 6.11. Boundary condition for sharp corner.

The equivalent formula of Laplace equation in finite difference form can be written as

$$h_{i,j} = (h_{i+1,j} + h_{i-1,j} + h_{i,j+1} + h_{i,j-1})/4 \tag{6.100}$$

In sharp corner boundary the following similar formula can be written as

$$h_{i,j} = \frac{1}{4}\left(h_{i+1,j} + h_{i-1,j}\right) + 2h_{i,j-1} \tag{6.101}$$

$$h_{i,j} = \frac{1}{4}\left(h_{i+1,j} + h_{i-1,j}\right) + 2h_{i+1,j} \tag{6.102}$$

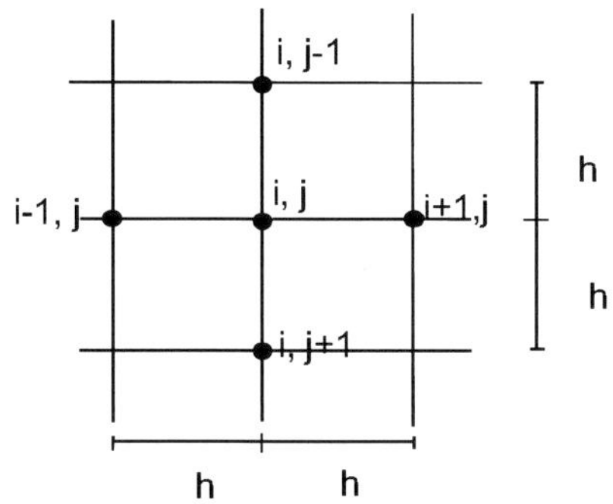

Figure 6.12 Labeling of grid points on flow domain.

Using the same principles some formulas can be written for some points in Figure 6.11,

At A: $h_{i,j} = \frac{1}{4}\left(h_{i+1,j} + h_{i-1,j}\right) + 2h_{i,j-1} = \frac{1}{1.5}\left(h_{i,j+1} + \frac{1}{2}\ h_{i-1,j}\right)$ (6.103)

At B: $h_{i,j} = \frac{1}{2}\left(h_{i-1,j} + h_{i,j+1}\right)$ (6.104)

At C: $h_{i,j} = \frac{1}{2.5}\left(h_{i,j+1} + h_{i-1,j} + \frac{1}{2}h_{i+1,j}\right)$ (6105)

At D and E: $h_{I,j} = \frac{1}{4}\left(h_{I+1,j} + h_{I-1,j}\right) + 2h_{I,j-1} = \frac{1}{2}\left(\frac{1}{2}h_{I+1,j} + \frac{1}{2}h_{I-1,j} + h_{I,j-1}\right)$ (6.106)

6.10. Initial Conditions

In terms of dependent variables, the initial condition describes the system's initial state. For instance, the simulation should begin with the definition of the initial water level and velocity in a channel network. Similar to this, it is necessary to establish the initial groundwater level in an aquifer system or region in order to build up the groundwater flow model.

Initial condition as one the physical meaning of the problem is required if the problem is time dependent. defines. The condition which describes the water flow of the beginning is the initial condition. A flow on a watershed can be given as an example to initial condition (See Figure 6.13):

The governing equation of the flow is defined by the continuity equation as,

$$\frac{\partial A}{\partial t} + \frac{\partial Q}{\partial x} = I \qquad (6.107)$$

where A is the cross-sectional area of the flow, Q is the discharge, and I is the rainfall excess.

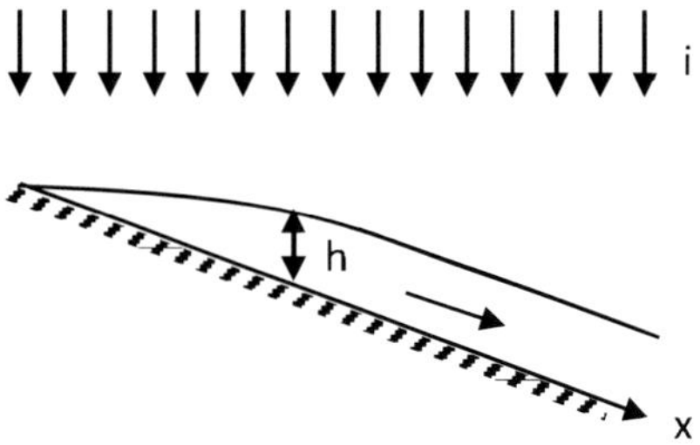

Figure 6.13 Surface flow definition in a watershed.

Initial condition for this problem can be written as:

$$Q(x, 0) = Q_0 \qquad (6.108)$$

$$A(x, 0) = A_0 \qquad (6.109)$$

Figure 6.14 shows inflow hydrograph with initial value of Q_0. If we use the kinematic model, the momentum equation can be written as,

$$Q = \alpha h^{\beta} \qquad (6.110)$$

Combining continuity and momentum equations yields:

$$\frac{\partial h}{\partial t} + \alpha h^{\beta-1}\frac{\partial h}{\partial x} = i \qquad (6.111)$$

Only h is an unknown value in this equation as the water depth, and I is the rainfall intensity as i = I/B, where B is the width.

Initial condition for this problem can be written for only h as:

$$h(x, 0) = 0.00 \tag{6.112}$$

Since there is no flow on upper part of the watershed, h will be zero at this section. So, the boundary condition can be written as,

$$h(0, t) = 0.00 \tag{6.113}$$

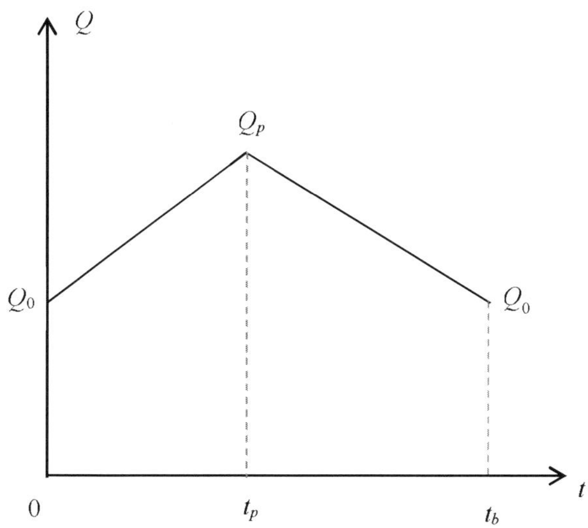

Figure 6.14 Example of inflow hydrograph.

6.11. Number of Boundary Conditions

Number of boundary conditions needed for solution for each variable is shown in Table 6.2.

In general, we can say that the number of boundary condition equals to half of the order of error.

As a result, we can modify the boundary conditions based on the physical circumstances within the hydraulic system, and when combined with the governing equation, we can attempt to solve the problem by employing these governing equations and boundary conditions. Therefore, they could either be partial or ordinary differential equations.

Table 6.2. Number of boundary conditions needed

The highest order derivative in the equation	Order of scheme used	Name of Scheme	Example	Number of points	Order of error	Number of boundary conditions for each variable
1	1	Forward	$\frac{\partial f}{\partial x}=\frac{f_{i+1,j}-f_{i,j}}{\Delta x}$	2	1	1
1	1	Backward	$\frac{\partial t}{\partial x}=\frac{f_{i,j}-f_{i-1,j}}{\Delta x}$	2	1	1
1	1	Centered	$\frac{\partial f}{\partial x}=\frac{f_{i+1,j}-f_{i-1,j}}{2\Delta x}$	3	2	1
1	2	Centered	$\frac{\partial f}{\partial x}=\frac{-f_{i+2,j}+8f_{i+1,j}-8f_{i-1,j}+f_{i-2,j}}{12\Delta x}$	5	4	2
2	1	Centered	$\frac{\partial^2 f}{\partial x^2}=\frac{f_{i+1,j}-2f_{i,j}+f_{i-1,j}}{\Delta x^2}$	3	2	1
2	2	Centered	$\frac{\partial^2 f}{\partial x^2}=\frac{-f_{i+2,j}+16f_{i+1,j}-30f_{i,j}+16f_{i+j}-f_{i-2,j}}{12\Delta x^2}$	5	4	2
3	2	Centered	$\frac{\partial^3 f}{\partial x^3}=\frac{-f_{i-2,j}+2f_{i-1,j}-2f_{i+1,j}+f_{i+1,j}}{2\Delta x^3}$	5	4	2
4	2	Centered	$\frac{\partial^4 f}{\partial x^4}=\frac{-f_{i-2,j}-4f_{i-1,j}+6f_{i,j}-4f_{i+1,j}}{2\Delta x^3}$	5	4	2

6.12. Applications of Boundary and Initial Conditions

6.12.1. Determination of Velocity Components from Non-Dimensional Stream Function

Find the velocity components at Point (3,5) for the flow domain which described as a horizontal sharp corner flow (see Figure 6.15).

Given data: B = 4 m, Q = 4 m^3/s, D = 1 m, $F_{i+1,,j} = F_{4,5} = 0.465$, $F_{i-1,j} = F_{2,5} = 0.560$, $F_{i,j+1} = F_{3,6} = 0.727$, $F_{i,j-1} = F_{3,4} = 0.315$.

Solution: The grid system is shown in Figure 6.16.

For Δx = 1 m, Δy = 1 m, Δx = Δy, we can calculate,

k = Q/(BD) = 4/(4x1) = 1.00

$\Delta x_1 = \Delta x/B = ¼ = 0.25$

$\Delta y_1 = \Delta y/B = ¼ = 0.25$

Velocity components can be calculated as,

$u = k(f_{i,j+1} - f_{i,j-1})/(2\Delta y_1) = 1.00x(0.727 - 0.315)/(2x0.25) = 0.824$ m/s

$v = -k(f_{i+1,j} - f_{i-1,j})/(2\Delta x_1) = -1.00x(0.465 - 0.560)/(2x0.25) = 0.190$ m/s

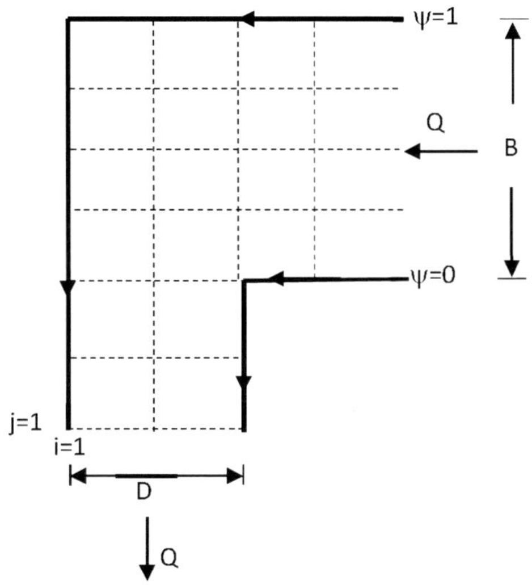

Figure 6.15. Flow domain.

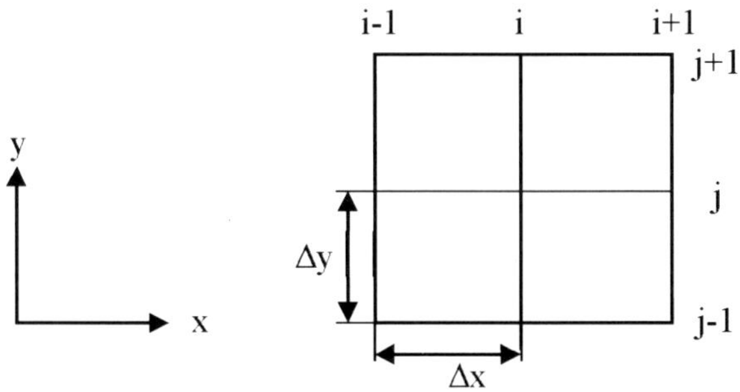

Figure 6.16. Grid system for finite difference approximation.

6.12.2. Example for Irregular Boundary Conditions

Find the Laplace equation for irregular boundary in an axial symmetric flow. The Laplace equation can be written in cylindrical coordinates as:

$$\frac{\partial^2 F}{\partial r^2} - \frac{1}{r}\frac{\partial F}{\partial r} + \frac{\partial^2 F}{\partial z^2} = 0 \tag{6.114}$$

We assume that the coordinate system starts from Point M, in Figure 6.17. The coordinates of some points can be written as; A(0,-sh), B(th, 0), C(0, h), D(-h, 0), where s and t are the ratios as MA/h = s, and MB/h = t, and h = Δr = Δz.

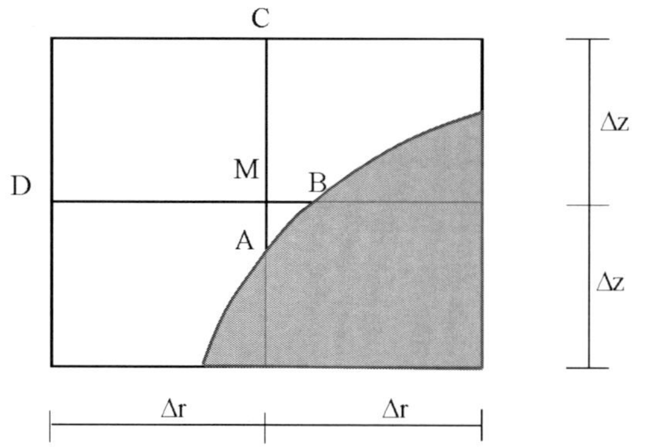

Figure 6.17. Irregular Boundary sketch.

The Taylor expansion can be written for at A, B, C, and D in Figure 6.17 as:

$$F_A = F_M - sh\left(\frac{\partial F}{\partial z}\right)_M + \frac{(sh)^2}{2}\left(\frac{\partial^2 F}{\partial z^2}\right)_M + \cdots \tag{6.115}$$

$$F_B = F_M + th\left(\frac{\partial F}{\partial r}\right)_M + \frac{(th)^2}{2}\left(\frac{\partial^2 F}{\partial r^2}\right)_M + \cdots \tag{6.116}$$

$$F_C = F_M + h\left(\frac{\partial F}{\partial z}\right)_M + \frac{(h)^2}{2}\left(\frac{\partial^2 F}{\partial z^2}\right)_M + \cdots \tag{6.117}$$

$$F_D = F_M - h\left(\frac{\partial F}{\partial r}\right)_M + \frac{(h)^2}{2}\left(\frac{\partial^2 F}{\partial r^2}\right)_M + \cdots \tag{6.118}$$

From those four equations one can obtain:

$$\left(\frac{\partial^2 F}{\partial z^2}\right)_M = \frac{2[F_A - F_M(1+s) + sF_c]}{h^2 s(1+s)} \tag{6.119}$$

$$\left(\frac{\partial^2 F}{\partial r^2}\right)_M = \frac{2[F_B - F_M(1+t) + tF_D]}{h^2 t(1+t)} \tag{6.120}$$

$$-\frac{1}{r}\left(\frac{\partial F}{\partial r}\right)_M = -\frac{1}{hr}\left[\frac{F_M(t-1)}{t} + \frac{F_B}{t(1+t)} - \frac{tF_D}{1+t}\right] \tag{6.121}$$

The Laplace equation for axial symmetric flow:

$$\left(\frac{\partial^2 F}{\partial r^2}\right)_M - \frac{1}{r}\left(\frac{\partial F}{\partial r}\right)_M + \left(\frac{\partial^2 F}{\partial z^2}\right)_M = \frac{2}{h^2}\left[\frac{F_A}{s(1+s)} + \frac{F_C}{1+s} - \frac{F_M}{s} - \frac{F_B}{t(1+t)} - \frac{F_M}{t}\right] - \frac{1}{hr}\left[\frac{F_M(t-1)}{t} + \frac{F_B}{t(1+t)} - \frac{tF_D}{1+t}\right] = 0 \tag{6.122}$$

Special cases: For s = 1, t = 1,

$$F_M = \left[\frac{2r(F_A + F_C)}{2} + \frac{2r(F_B + F_D) - h(F_B - F_D)}{2}\right]\left[\frac{1}{2r + 2r + h - h}\right] \tag{6.123}$$

$$F_M = \left[\frac{F_A + F_B + F_C + F_D}{4}\right] - \frac{1}{8r}(F_B - F_D) \tag{6.124}$$

For Cartesian coordinates (for $r = \infty$)

$$F_M = \frac{F_A + F_B + F_C + F_D}{4} \tag{6.125}$$

From Equation (6.122), the general form of the equivalent Laplace equation can be written as

$$F_M = \left[\frac{2rt(F_A + sF_C)}{1+s} + \frac{2rs(F_B + tF_D) - hs(F_B - t^2 F_D)}{1+t}\right]\left[\frac{1}{2rt + 2rs + sht - sh}\right] \tag{6.126}$$

6.12.3. Determination of Boundary Conditions for Navier-Stokes Equations in Vorticity/Stream Function Form

Consider the classical Lid-Driven Cavity Problem using the Navier-Stokes equations in vorticity form.

a) Derive the Navier-Stokes Equations in Vorticity/Stream Function form.
b) Determine Boundary Conditions.

Solution: (a) Derive the Navier-Stokes Equations in Vorticity/Stream Function form.

We can examine the flow in the cavity filled with a fluid as shown in Figure 6.18.

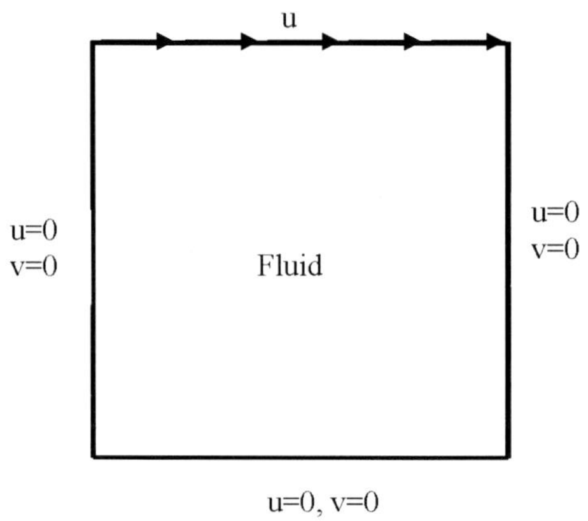

Figure 6. 18 Schematic of the lid-cavity problem.

We can write the Navier –Stokes equation in incompressible two-dimensional flow. Continuity equation can be written as Equation (6.1). We can write the momentum equations as in Equation (6.2) and Equation (6.3).

The Vorticity Equation: Deriving the Equation (6.2) by $\partial/\partial y$, and the Equation (6.3) by $\partial/\partial x$ and adding these new two equations yields the vorticity equation as:

$$\frac{\partial \xi}{\partial t} + u\frac{\partial \xi}{\partial x} + v\frac{\partial \xi}{\partial y} = \frac{1}{R_e}\left(\frac{\partial^2 \xi}{\partial x^2} + \frac{\partial^2 \xi}{\partial y^2}\right) \tag{6.127}$$

where vorticity is defined as

$$\xi = \frac{\partial v}{\partial x} - \frac{\partial u}{\partial y} \tag{6.128}$$

The Stream Function Equation: We define the stream function, ψ, as in Equation (6.12) and Equation (6.13), which automatically satisfies the continuity equation, Equation (6.1) as,

$$\frac{\partial}{\partial x}\left(\frac{\partial \psi}{\partial y}\right) - \frac{\partial}{\partial y}\left(\frac{\partial \psi}{\partial x}\right) = 0 \tag{6.129}$$

Substituting the Equation (6.127) and Equation (6.128) into the definition of the vorticity, Equation (6.1), yields Equation (6.7). This is an elliptic equation which is called Poisson's equation $\partial^2 \psi/\partial x^2 + \partial^2\psi/\partial y^2 = -\xi$.

As a summary, the Navier-Stokes equations in vorticity-stream function form are:

Advection/diffusion equation, Equation (6.128), and Poisson equation, Equation (6.7), as $\partial^2 \psi/\partial x^2 + \partial^2\psi/\partial y^2 = -\xi$.

(b) Boundary Conditions: We can write the boundary conditions of the examined problem for velocity components as can be seen in Figure 6.18.

Boundary Conditions for Stream Function: Boundary conditions for the stream function can be written as: At the right and the left boundary: For $u = 0 \Rightarrow \partial\psi/\partial y = 0$,

$$\psi = \text{Constant} \tag{6.130}$$

At the top and the bottom boundary: For $v = 0 \Rightarrow \partial\psi/\partial x = 0$, $\psi =$ Constant.

The constant must be the same on all boundaries because the boundaries meet, $\psi =$ Constant.

Boundary Conditions for the Vorticity: As can be seen from Figure 6.7, the stream function is a constant on the wall, hence the normal velocity is zero. However, the zero tangential velocity needs to be enforced.

At the right and left boundary: For $v = 0$,

$$\frac{\partial\psi}{\partial x} = 0 \tag{6.131}$$

At the bottom boundary: For $u = 0$,

$$\frac{\partial\psi}{\partial y} = 0 \tag{6.132}$$

At the top boundary: For $u = U_{wall}$, $\partial\psi/\partial y = 0$.

The stream function must be used to determine the wall vorticity. On the walls, the stream function is constant.

At the right and the left boundary: For Poisson Equation, $\partial^2 \psi/\partial x^2 + \partial^2\psi/\partial y^2 = -\xi$, we can write,

$$\xi_{wall} = -\frac{\partial^2\psi}{\partial x^2} \tag{6.133}$$

In a similar manner, at the top and bottom boundaries: For $\partial^2\psi/\partial x^2 + \partial^2\psi/\partial y^2 = -\xi$, we can write,

$$\xi_{wall} = -\frac{\partial^2 \psi}{\partial y^2} \tag{6.134}$$

Chapter 7

Solutions of Elliptic Equations

7.1. Introduction

Elliptic partial differential equations are used in the representation of a wide range of physics events. In engineering, besides hydraulic engineering, they have wide applications in structural engineering, coastal engineering, mechanical engineering, and electrical engineering. Some hydraulic engineering problems are also defined by elliptic equations as potential flows in pipes and open channels. This list can be extended further.

In this chapter, the solution of elliptic partial differential equations using finite difference method will be studied and application to various problems will be achieved. Elliptic partial differential equations can be described as boundary value problems. Their solution domains are closed. These problems are called equilibrium state and are known as steady state case. These equations do not contain any time term. However, like other type of partial differential equations involving time, they examine two-dimensional spaces as x and y variables in space. Because of these two-dimensionality, some of the methods used in the other type of differential equations, can also be used in solution of elliptic partial differential equations. Using the finite difference method by the discretized system, an elliptic partial differential equation turns into a set of algebraic equations (Lynch, 2005; Fenton, 2010; Mohan, 2010).

7.2. Governing Equations

In hydraulics and fluid mechanics there are some examples of elliptic partial differential equations, the most common occurrence of elliptic partial differential equations for us is steady state diffusion problems. We can write non-steady state diffusion equation in two dimensions as,

$$D_x \frac{\partial^2 u}{\partial x^2} + D_y \frac{\partial^2 u}{\partial y^2} = \frac{\partial u}{\partial t} \tag{7.1}$$

At steady state, $i.e., \frac{\partial u}{\partial t} = 0$, Equation (7.1) becomes:

$$D_x \frac{\partial^2 u}{\partial x^2} + D_y \frac{\partial^2 u}{\partial y^2} = 0 \tag{7.2}$$

This is the steady sate diffusion equation. This equation is always elliptic for all values of D_x and D_y. If coefficients D_x and D_y is equal to 1, we can obtain the Laplace equation as:

$$\frac{\partial^2 u}{\partial x^2} + \frac{\partial^2 u}{\partial y^2} = 0 \tag{7.3}$$

We can also write the Poisson equation as

$$\frac{\partial^2 u}{\partial x^2} + \frac{\partial^2 u}{\partial y^2} = g(x, y) \tag{7.4}$$

7.3. Finite Difference Formulation of Laplace Equation

In boundary value problems, the function u(x, y) is generally known at all boundary points of a rectangular domain. In such cases, the finite difference method is commonly used to solve governing equations.

We consider second order analogs for $\partial^2 u/\partial y^2$ and $\partial^2 u/\partial x^2$ for Laplace Equation, Equation (7.3):

$$\frac{u_{i+1,j} - 2u_{i,j} + u_{i-1,j}}{(\Delta x)^2} + \frac{u_{i,j+1} - 2u_{i,j} + u_{i,j-1}}{(\Delta y)^2} = 0 \tag{7.5}$$

For convenience, we can take $\Delta x = \Delta y$,

$$u_{i,j} = (u_{i+1,j} + u_{i-1,j} + u_{i,j+1} + u_{i,j-1})/4 \tag{7.6}$$

A five-point stencil for elliptic analog corresponding to Equation 7.6 is shown in Figure 7.1.

A partial differential equation together with the given boundary condition is well posed when: 1. A solution exists, 2. The solution is unique, 3. The solution depends on the data. Boundary conditions are very important.

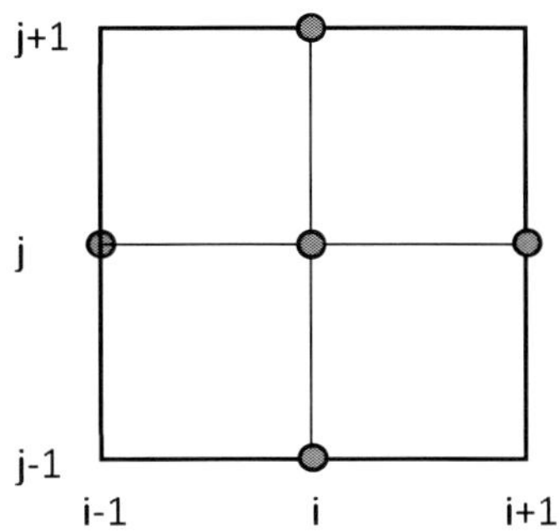

Figure 7.1. A five-point stencil for elliptic partial differential equations.

The solution of elliptic partial differential equations depends on the boundary conditions (Stelling and Booij, 1999).

Generally, there are three most common boundary conditions as:

- Dirichlet type boundary conditions. In this case, the *f* values are known on the boundaries of the solution domain.
- Neumann type boundary conditions. In some cases, the gradient of *f* normal to the boundaries is known together with a value of *f*, $(\partial f/\partial n)$.
- Mixed boundary conditions. In some cases, some *f* values and some gradient values are known around the solution domain.

A problem related elliptic equations are called boundary value problem including closed domain with boundary conditions, as sketched in Figure 7.2. The function values at boundary points are generally known.

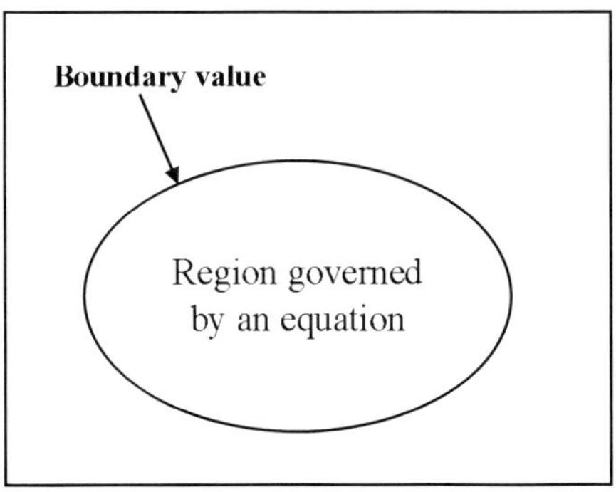

Figure 7.2. Definition sketch for boundary value problems.

7.4. Solution Methods for Elliptic Partial Differential Equations

To discuss the solution methods for elliptic partial differential equations we will take the Laplace equation into consideration,

$$\frac{\partial^2\phi}{\partial x^2}+\frac{\partial^2\phi}{\partial y^2}=0 \tag{7.7}$$

In the solution of elliptic partial differential equations, it is common to use the central finite difference approximation for second order derivatives. It is possible to have an implicit equation when this approximation is employed.

We can express $\partial^2\phi/\partial x^2$ and $\partial^2\phi/\partial y^2$ for the first order using the four adjacent mesh points as can be seen in Figure 7.3 to obtain the finite difference expression as:

$$\frac{\phi_{I+1,j}-2\phi_{I,j}+\phi_{I-1,j}}{\Delta x^2}+\frac{\phi_{I,j+1}-2\phi_{I,j}+\phi_{I,j-1}}{\Delta y^2}=0 \tag{7.8}$$

By selecting the equal space step in space discretization as square mesh, $\Delta x = \Delta y$, Equation (7.8) becomes as,

$$\frac{\phi_{i+1,j}+\phi_{i-1,j}+\phi_{i,j+1}+\phi_{i,j-1}-4\phi_{i,j}}{\Delta x^2}=0 \tag{7.9}$$

In the i, j point, f values are taken as $\phi_{i,j}$. Therefore, for all the nodal points in the solution domain, the difference equation, Equation (7.9), are coupled as a number of nodal points and solved simultaneously.

We also need boundary conditions for the solution domain. Dirichlet, von Neumann or mixed boundary conditions can specify the boundary values of *i,j*. Equation (7.9) in combination with the boundary conditions may be solved in a variety of solution methods (Koutitas, 1983).

The set of algebraic equations found has a tri-diagonal coefficient matrix. Five common methods are used to solve this set of equations, often called relaxation techniques or iterative methods (McDonough, 2008). These are:

a) Direct solution method using standard Gauss elimination,
b) Jacobi iteration,

c) Gauss-Seidel iteration,
d) Successive Over Relaxation (SOR) method,
e) Line Successive Over Relaxation (SLOR) method.

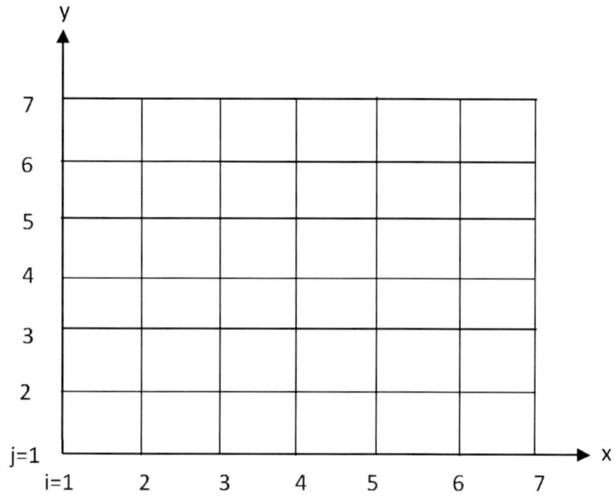

Figure 7.3. Computational grid for elliptic equation.

7.4.1. Direct Method

If the boundary conditions are linear, the finite difference approach is also linear and Gauss Elimination approach can be used to solve the resulting set of equations. If the total number of nodes required (m + 1) (n + 1) is relatively small, this approach can be applied. Since, the matrix A used to represent the entire system will have $[(m + 1) (n + 1)]^2$ elements, in such a solution, computer storage and computing costs cannot be suitable even for relatively small m and n numbers.

The structure of the system ensures that the matrix A is relatively sparse. Thus, the matrix consists of three diagonal core, one of which is non-zero diagonal and the other is diagonal below it. These nonzero diagonals are shifted from the leading diagonal by m or n. The pivoting provided (if necessary) is performed so as not to place any non-zero elements outside this band. Then the solution is found with standard Gauss elimination. By producing only non-zero elements in this band, storage and calculation requirements are significantly reduced. Careful selection of the order of the matrix elements can help reduce the size of this matrix. Thus, for a square area, it should contain only small elements.

Special solvers have been developed for this problem, since the need to solve the Laplace equation and other related equations is widely increased. One of the best of these is Hockney's method. This method solves Ax = b, which can be used to reduce the block tridiagonal matrix and its corresponding right side.

The form of this matrix reduces,

$$A = \begin{bmatrix} \hat{A} & I & & & & \\ I & \hat{A} & I & & & \\ & I & \hat{A} & I & & \\ & & I & \hat{A} & \ddots & \\ & & & \ddots & \ddots & I \\ & & & & I & \hat{A} \end{bmatrix} \tag{7.10}$$

into a block diagonal matrix of the form as,

$$A = \begin{bmatrix} \hat{B} & & & & & \\ & \hat{B} & & & & \\ & & \hat{B} & & & \\ & & & \hat{B} & & \\ & & & & \ddots & \\ & & & & & \hat{B} \end{bmatrix} \tag{7.11}$$

where $\hat{A}$ and $\hat{B}$ are block tri-diagonal matrices I is an identity matrix. This can be done iteratively to reduce the n-dimensional finite difference approximation of the equation to Laplace equation to the tri-dimensional system of equations with n-1 applications. Calculation cost is very low. The main disadvantage of this method is that boundary conditions must be able to throw into block tridiagonal format.

In the solution of elliptic partial differential equations, another alternative to the direct solution of finite difference equations is an iterative numerical solution. These iterative methods are often called relaxation methods, allowing the first guess in the solution to slowly relax to the actual solution. This way reduces errors. There are several approaches with different complexity and speed. These methods will be introduced below.

7.4.2. Jacobi Iteration

The Jacobi Iteration is a simple method. In order to find the solution for a two-dimensional Laplace equation, considering Figure 7.4, we can take the following steps:

1. We estimate some initial values for $\phi_{i,j}$.
2. We define the boundary conditions.
3. We calculate ϕ^*_{ij} for each internal mesh point as

$$\phi^*_{ij} = (\phi_{i+1,j} + \phi_{i-1,j} + \phi_{i,j+1} + \phi_{i,j-1})/4 \tag{7.12}$$

4. We replace old solution ϕ with new estimate ϕ^*.
5. If solution does not satisfy the predicted tolerance, we repeat from step 2.

The coefficients in Equation (7.12) are shown Figure 7.5, which is often referred to as the *stencil.*

Although every iteration in Jacobi iteration is simple and inexpensive, this iteration is very slow to converge, especially if there are larger grids.

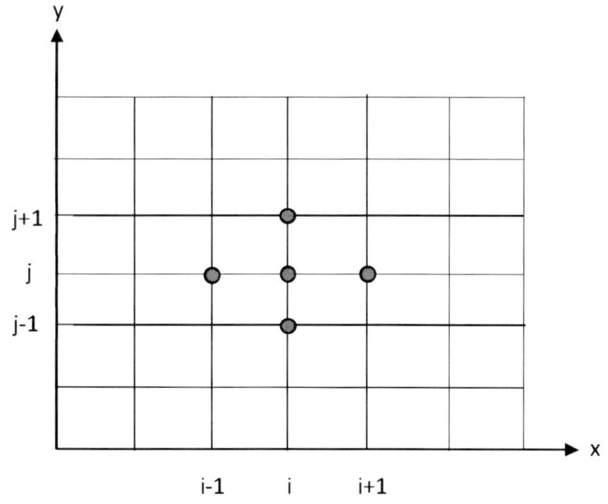

Figure 7.4. Five-point stencil for calculation of Laplace equation.

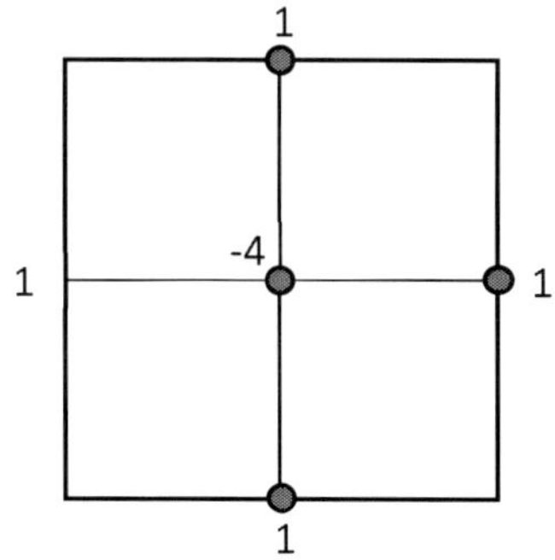

Figure 7.5 Stencil coefficients for finite difference in Laplace equation.

7.4.3. Gauss-Seidel Iteration

In fact, the Gauss-Seidel Recurrence is very similar to the Jacobi Recurrence. The only difference is that the new $\phi^*_{i,j}$ estimate allows it to be returned to the solution $\phi_{i,j}$ as soon as it is completed and allows its use to be used immediately rather than postponing it to the next iteration. The advantages of this iteration are:

1. ϕ^* requires less memory as it does not need to be stored in computer.
2. Although the calculation is still relatively slow, it converges faster.

On the other hand, the method is less suitable for vectorization for a given iteration. The new estimate of a mesh point is based on new estimates for those already scanned.

7.4.4. Successive Over Relaxation Method

The method of successive over-relaxation, SOR, is an alternative method for solving an elliptic partial differential equation, as an example the Laplace equation. This method yields results in faster convergence.

In the solution found by any of the three methods previously described, it was obtained that the errors only decrease slowly and decrease in a monotonous manner. For this reason, instead of Equation (7.12), Equation (7.13) is used in the SOR method as,

$$\phi^*_{i,j} = (1\text{-}s)\phi_{i,j} + s(\phi_{i+1,j} + \phi_{i-1,j} + \phi_{i,j+1} + \phi_{i,j-1})/4 \quad (7.13)$$

where s is a coefficient.

The optimal value of s depends on the problem being solved. The value of s can vary as the iteration process converges. However, a typical value of s around 1.2 to 1.8 gives good results. In some special cases it is possible to determine an optimum value of s analytically.

In order to solve the Laplace equation in a 2-D domain, generally, the following five steps are taken as

1. We discretize domain into grid of evenly spaced points
2. In the flow domain for all nodes where ϕ is unknown:

$\Delta x = \Delta y = h$, substitute into main equation

3. We consider the boundary conditions, and we write, n*m equations for

$u(x_i = 1{:}m,\ y_j = 1{:}n)$ or n*m unknowns.

4. We solve the set of algebraic equations, n*m equations, simultaneously with an efficient scheme using any iteration method.

7.5. Applications of Elliptic Partial Differential Equations

7.5.1. Numerical Solution to the Groundwater Flow Equation

For a homogeneous and isotropic aquifer, the final values of potential head at four grids are given as:

$\phi_{i+1,j} = 1.8;\ \phi_{i-1,j} = 1.60;\ \phi_{i,j+1} = 1.9;\ \phi_{i,j-1} = 1.5.$

Compute the potential head at grid (i, j) considering steady flow and no withdrawal from the grid.

Solution: The grid point system for finite difference method is shown in Figure 7.3.

Let us assume that we are interested in solving the Laplace equation in 2-dimensions as in Equation (7.7), where ϕ is the velocity potential function.

We substitute the second order partial derivatives with their finite difference forms, and setting for simplicity, $\Delta x = \Delta y$, at the Laplace equation, we obtain:

$$\phi_{i,j} = (\phi_{i+1,j} + \phi_{i-1,j} + \phi_{i,j+1} + \phi_{i,j-1})/4 \quad (7.14)$$

which is the finite difference expression of the Laplace equation.

Using the given data, we obtain,

$$\phi_{i,j} = (\phi_{i+1,j} + \phi_{i-1,j} + \phi_{i,j+1} + \phi_{i,j-1})/4 = (1.8 + 1.6 + 1.9 + 1.5)/4 = 1.7$$

7.5.2. Numerical Solution of Stream Function Values, Velocities, Flow Rate for Flow, Potential Function Values and Pressures under Sheet Piles

A riverbed is made up of 8.5 m thick sand on an impermeable rock layer. The water depth is 2.5 m. A 4.0 m wide long cofferdam will be created for construction. For this purpose, two sheet piles, which are 6 m below the riverbed, will be used. As shown in Figure 7.6, the water level within the cofferdam will be excavated at a depth of 2 m. The water level in the cofferdam will be kept at the excavation level by pumping.

a. Find the stream function values in the flow domain.
b. Find the velocity components under of sheet piles, if the hydraulic conductivity of the soil is 7.0x 10^{-4} m/s.
c. Find the flow rate of the pump ($m^3/s/m$) for trench, if the water level within the sheet piles was maintained at the bottom of the trench.
d. Find the potential function values in the flow domain.
e. Find pressures at point for each vertical distance from soil on and back of sheet pile.

Solution: a. Find the stream function values in the flow domain.

Flow under two sheet piles is a symmetric flow as seen in Figure 7.6. So, we can consider the half of the domain.

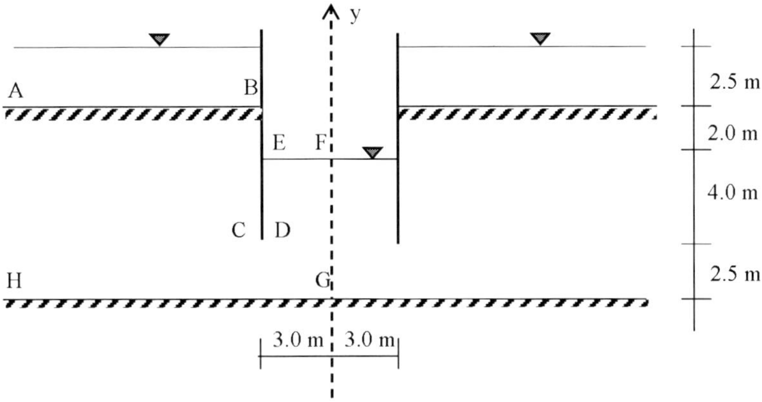

Figure 7.6. Dimensions of flow domain with two impervious sheet piles.

1. Assuming the flow is a potential flow, we can solve the Laplace Equation.
2. Using a finite difference method, for a square mesh ($\Delta x = \Delta y = 0.5$ m), the equivalent of Laplace equation can be written for stream function as

$$\psi_{i,j} = (\psi_{i+1,j} + \psi_{i-1,j} + \psi_{i,j+1} + \psi_{i,j-1})/4$$

3. For convenience, the dimensionless form of stream function is choosing.
4. Boundary conditions for domain ABCDEFGH which has (20x18) grid points, (See Figure 7.6) are defined as follows.

Solid walls, as sheet pile: on BC, CD, DE:

$$\psi_{i,,j} = 1.0$$

Impervious layer and symmetry axis: on HG, and GF:

$$\psi_{i,j} = 0.0$$

Upper part of flow domain: on AB

$$\frac{\partial \psi}{\partial y} = 0$$

Water level in the cofferdam, on EF:

$$\frac{\partial \psi}{\partial y} = 0$$

Upstream of domain: assuming a linear distribution for stream function, on AH:

$$\psi_{i,j} = 2\psi_{i+1}j - \psi i_{+2,j}$$

Boundary conditions and first estimation for interior points are shown in Figure 7.7.

0.50	0.50	0.50	0.50	0.50	0.50	0.50	0.50	0.50	0.50	0.50	0.50	1.00							
0.50	0.50	0.50	0.50	0.50	0.50	0.50	0.50	0.50	0.50	0.50	0.50	1.00							
0.50	0.50	0.50	0.50	0.50	0.50	0.50	0.50	0.50	0.50	0.50	0.50	1.00							
0.50	0.50	0.50	0.50	0.50	0.50	0.50	0.50	0.50	0.50	0.50	0.50	1.00							
0.50	0.50	0.50	0.50	0.50	0.50	0.50	0.50	0.50	0.50	0.50	0.50	1.00	1.00	0.50	0.50	0.50	0.50	0.50	0.00
0.50	0.50	0.50	0.50	0.50	0.50	0.50	0.50	0.50	0.50	0.50	0.50	1.00	1.00	0.50	0.50	0.50	0.50	0.50	0.00
0.50	0.50	0.50	0.50	0.50	0.50	0.50	0.50	0.50	0.50	0.50	0.50	1.00	1.00	0.50	0.50	0.50	0.50	0.50	0.00
0.50	0.50	0.50	0.50	0.50	0.50	0.50	0.50	0.50	0.50	0.50	0.50	1.00	1.00	0.50	0.50	0.50	0.50	0.50	0.00
0.50	0.50	0.50	0.50	0.50	0.50	0.50	0.50	0.50	0.50	0.50	0.50	1.00	1.00	0.50	0.50	0.50	0.50	0.50	0.00
0.50	0.50	0.50	0.50	0.50	0.50	0.50	0.50	0.50	0.50	0.50	0.50	1.00	1.00	0.50	0.50	0.50	0.50	0.50	0.00
0.50	0.50	0.50	0.50	0.50	0.50	0.50	0.50	0.50	0.50	0.50	0.50	1.00	1.00	0.50	0.50	0.50	0.50	0.50	0.00
0.50	0.50	0.50	0.50	0.50	0.50	0.50	0.50	0.50	0.50	0.50	0.50	1.00	1.00	0.50	0.50	0.50	0.50	0.50	0.00
0.50	0.50	0.50	0.50	0.50	0.50	0.50	0.50	0.50	0.50	0.50	0.50	1.00	1.00	0.50	0.50	0.50	0.50	0.50	0.00
0.50	0.50	0.50	0.50	0.50	0.50	0.50	0.50	0.50	0.50	0.50	0.50	0.50	0.50	0.50	0.50	0.50	0.50	0.50	0.00
0.50	0.50	0.50	0.50	0.50	0.50	0.50	0.50	0.50	0.50	0.50	0.50	0.50	0.50	0.50	0.50	0.50	0.50	0.50	0.00
0.50	0.50	0.50	0.50	0.50	0.50	0.50	0.50	0.50	0.50	0.50	0.50	0.50	0.50	0.50	0.50	0.50	0.50	0.50	0.00
0.50	0.50	0.50	0.50	0.50	0.50	0.50	0.50	0.50	0.50	0.50	0.50	0.50	0.50	0.50	0.50	0.50	0.50	0.50	0.00
0.00	0.00	0.00	0.00	0.00	0.00	0.00	0.00	0.00	0.00	0.00	0.00	0.00	0.00	0.00	0.00	0.00	0.00	0.00	0.00

Figure 7.7. First estimated values for stream function.

Computed dimensionless stream function values for after 100 iterations are shown in Figure 7.8, using $\psi_{i,j} = (\psi_{i+1,j} + \psi_{i-1,j} + \psi_{i,j+1} + \psi_{i,j-1})/4$.

Find the velocity components under of sheet piles, if the hydraulic conductivity of the soil is 7.0x 10^{-4} m/s.

Velocity components: From definitions for stream function we can write

$$u = \frac{\partial \psi}{\partial y} = k\frac{\partial F}{\partial y_1}$$

$$v = -\frac{\partial \psi}{\partial x} = -k\frac{\partial F}{\partial x_1}$$

where F is non-dimensional stream function as defined $F = \psi/\psi_{max}$. The equivalent of these can be written in finite difference form as

$$u_{i,j} = k\frac{F_{i,j+1} - F_{i,j-1}}{2\Delta y_1}$$

$$v_{i,j} = -k\frac{F_{i+1,j} - F_{i-1,j}}{2\Delta x_1}$$

where $\Delta x_1 = \Delta x/B$, and $\Delta y_1 = \Delta y/B$.

0.31	0.36	0.41	0.46	0.51	0.57	0.62	0.68	0.74	0.81	0.87	0.93	1.00							
0.31	0.36	0.41	0.46	0.51	0.57	0.62	0.68	0.74	0.81	0.87	0.93	1.00							
0.31	0.35	0.40	0.46	0.51	0.56	0.62	0.68	0.74	0.80	0.87	0.93	1.00							
0.30	0.35	0.40	0.45	0.50	0.56	0.61	0.67	0.74	0.80	0.86	0.93	1.00							
0.29	0.34	0.39	0.44	0.49	0.55	0.60	0.66	0.73	0.79	0.86	0.93	1.00	1.00	0.83	0.66	0.49	0.33	0.16	0.00
0.28	0.32	0.37	0.42	0.48	0.53	0.59	0.65	0.72	0.78	0.85	0.93	1.00	1.00	0.83	0.66	0.49	0.33	0.16	0.00
0.26	0.31	0.36	0.41	0.46	0.51	0.57	0.64	0.70	0.77	0.85	0.92	1.00	1.00	0.83	0.66	0.49	0.33	0.16	0.00
0.24	0.29	0.34	0.39	0.44	0.49	0.55	0.61	0.68	0.76	0.83	0.92	1.00	1.00	0.83	0.66	0.49	0.32	0.16	0.00
0.22	0.27	0.31	0.36	0.41	0.47	0.52	0.59	0.66	0.73	0.82	0.91	1.00	1.00	0.82	0.65	0.48	0.32	0.16	0.00
0.20	0.24	0.29	0.33	0.38	0.43	0.49	0.55	0.62	0.70	0.79	0.89	1.00	1.00	0.82	0.64	0.47	0.31	0.15	0.00
0.18	0.22	0.26	0.30	0.35	0.40	0.45	0.51	0.58	0.66	0.76	0.87	1.00	1.00	0.80	0.62	0.45	0.29	0.15	0.00
0.15	0.29	0.23	0.27	0.31	0.35	0.41	0.46	0.53	0.61	0.71	0.84	1.00	1.00	0.77	0.58	0.42	0.27	0.13	0.00
0.13	0.26	0.19	0.23	0.27	0.31	0.35	0.40	0.46	0.54	0.63	0.77	1.00	1.00	0.71	0.52	0.37	0.24	0.12	0.00
0.10	0.13	0.16	0.19	0.22	0.25	0.29	0.33	0.38	0.44	0.51	0.60	0.68	0.67	0.55	0.42	0.31	0.20	0.10	0.00
0.08	0.10	0.12	0.14	0.17	0.19	0.22	0.25	0.29	0.34	0.38	0.43	0.46	0.45	0.39	0.31	0.23	0.15	0.07	0.00
0.05	0.07	0.08	0.10	0.11	0.13	0.15	0.17	0.20	0.22	0.25	0.28	0.29	0.28	0.25	0.20	0.15	0.10	0.05	0.00
0.03	0.03	0.04	0.05	0.06	0.07	0.08	0.09	0.10	0.11	0.13	0.14	0.14	0.14	0.12	0.10	0.08	0.05	0.03	0.00
0.00	0.00	0.00	0.00	0.00	0.00	0.00	0.00	0.00	0.00	0.00	0.00	0.00	0.00	0.00	0.00	0.00	0.00	0.00	0.00

Figure 7.8. Computed non- dimensional stream function values.

Using the above formulation for k $=7.0\text{x}10^{-4}$ m/s, $\Delta x = \Delta y = 0.5$ m and B = 2.5 m, the horizontal components of the velocity under the sheet pile can be obtain,

$u_{13,6} = 0.00$ m/s

$u_{13,5} = 17.5\text{x}10^{-4}$ m/s x(1.00 - 0.46) = $9.45\text{x}10^{-4}$ m/s

$u_{13,4} = 17.5\ \text{x}10^{-4}$ m/s x(0.68 - 0.29) = $6.83\ \text{x}10^{-4}$ m/s

$u_{13,3} = 17.5\ \text{x}10^{-4}$ m/s x(0.46 - 0.14) = $5.60\text{x}10^{-4}$ m/s

$u_{13,2} = 17.5\ \text{x}10^{-4}$ m/s x(0.29 - 0.00) = $5.08\text{x}10^{-4}$ m/s

$u_{13,1} = 0.00$ m/s.

Similarly, we can calculate the vertical components of velocity under the sheet pile as

$v_{13,6} = 0.00$ m/s

$v_{13,5} = 17.5 \times 10^{-4}$ m/s x(0.67- 0.60) = 1.23×10^{-4} m/s

$v_{13,4} = 17.5 \times 10^{-4}$ m/s x(0.45 - 0.43) = 0.35×10^{-4} m/s

$v_{13,3} = 17.5 \times 10^{-4}$ m/s x0.0 = 0.00 m/s

$v_{13,2} = 0.00$ m/s

$v_{13,1} = 0.00$ m/s.

Resultant velocities: one can obtain using $V = (u^2 + v^2)^{0,5}$

$V_{13,6} = 0.00$ m/s

$V_{13,5} = 9.53 \times 10^{-4}$ m/s

$V_{13,4} = 6.84 \times 10^{-4}$ m/s

$V_{13,3} = 5.60 \times 10^{-4}$ m/s

$V_{13,2} = 5.08 \times 10^{-4}$ m/s

$V_{13,1} = 0.00$ m/s.

c. Find the flow rate of the pump (m^3/s/m) for trench, if the water level within the sheet piles was maintained at the bottom of the trench.

Flow rate per unit length using continuity equation as (Q =VA), q = Q/D $= \sum_{j=1}^{j=5} \Delta y \, (V_{i,j} + V_{i,j+1})/2$, can be calculated for $\Delta y = 0.5$ m:

q = 0.5[(0.00 + 9.53)/2 + (9.53 + 6.84)/2 + (6.84 + 5.60)/2 + (5.60 + 5.08)/2 + (5.08 + 0.00)/2] $\times 10^{-4} = 13.53 \times 10^{-4}$ m^3/s/m

or

$$q = 4.89 \text{ m}^3/\text{hour/m}.$$

In calculation, we consider only the half of the domain, So, the flow rate for all domain is obtain as

Flow rate = 2xq = 9.74 m^3/hour/m.

d. Find the potential function values in the flow domain.

Assuming the flow is a potential flow, we can employ the Laplace Equation as the governing equation.

Using FDM, for a square mesh ($\Delta x = \Delta y = 0.5$ m), the equivalent of Laplace equation can be written for potential function as $\phi_{i,j} = (\phi_{i+1,j} + \phi_{i-1,j} + \phi_{i,j+1} + \phi_{i,j-1})/4$

For convenience, the dimensionless form of potential function is chosen.

Boundary conditions for domain ABCDEFGH which has (20x18) grid points, (See Figure 7.6) are defined as follows.

Solid walls, such as sheet pile: on BC, CD, DE:

$$\frac{\partial \phi}{\partial x} = 0$$

Impervious layer and symmetry axis: on HG, and GF:

$$\frac{\partial \phi}{\partial x} = 0$$

$$\frac{\partial \phi}{\partial y} = 0$$

Upper part of flow domain: on AB

$$\phi = 2.5$$

Water level in the cofferdam, on EF:

$$\phi = 0$$

Upstream of domain: assuming a linear distribution for stream function, on AH:

$$\frac{\partial \phi}{\partial y} = 0$$

Figure 7.9 depicts dimensionless stream function values computed by Microsoft Excel after 100 iterations, using the formulation of $\phi_{i,j} = (\phi_{i+1,j} + \phi_{i-1,j} + \phi_{i,j+1} + \phi_{i,j-1})/4$ and boundary conditions.

	i=1	2	3	4	5	6	7	8	9	10	11	12	13	14	15	16	17	18	19	20
18	1.00	1.00	1.00	1.00	1.00	1.00	1.00	1.00	1.00	1.00	1.00	1.00	1.00							
17	0.99	0.99	0.99	0.99	0.99	0.99	0.99	0.99	0.99	0.99	0.98	0.98	0.98							
16	0.97	0.97	0.97	0.97	0.97	0.97	0.97	0.97	0.97	0.97	0.97	0.97	0.97							
15	0.96	0.96	0.96	0.96	0.96	0.96	0.96	0.96	0.96	0.95	0.95	0.95	0.95							
14	0.95	0.95	0.95	0.95	0.95	0.94	0.94	0.94	0.94	0.94	0.94	0.94	0.94	0.00	0.00	0.00	0.00	0.00	0.00	0.00
13	0.93	0.93	0.93	0.93	0.93	0.93	0.93	0.93	0.92	0.92	0.92	0.92	0.92	0.04	0.04	0.04	0.04	0.04	0.04	0.04
12	0.92	0.92	0.92	0.92	0.92	0.92	0.91	0.91	0.91	0.91	0.90	0.90	0.90	0.09	0.83	0.66	0.49	0.33	0.16	0.09
11	0.91	0.91	0.91	0.91	0.90	0.90	0.90	0.90	0.89	0.89	0.89	0.89	0.89	0.13	0.83	0.66	0.49	0.32	0.16	0.13
10	0.90	0.90	0.90	0.89	0.89	0.89	0.88	0.88	0.87	0.87	0.87	0.87	0.87	0.18	0.82	0.65	0.48	0.32	0.16	0.18
9	0.89	0.88	0.88	0.88	0.88	0.87	0.87	0.86	0.86	0.85	0.85	0.84	0.84	0.22	0.82	0.64	0.47	0.31	0.15	0.22
8	0.87	0.87	0.87	0.87	0.87	0.86	0.85	0.85	0.84	0.83	0.82	0.82	0.82	0.27	0.80	0.62	0.45	0.29	0.15	0.26
7	0.86	0.86	0.86	0.86	0.85	0.85	0.84	0.83	0.82	0.81	0.80	0.79	0.79	0.32	0.77	0.58	0.42	0.27	0.13	0.30
6	0.86	0.85	0.85	0.85	0.84	0.84	0.83	0.81	0.80	0.78	0.77	0.75	0.75	0.37	0.71	0.52	0.37	0.24	0.12	0.33
5	0.85	0.85	0.85	0.84	0.83	0.82	0.81	0.80	0.78	0.76	0.73	0.70	0.61	0.50	0.55	0.42	0.31	0.20	0.10	0.37
4	0.84	0.84	0.84	0.83	0.83	0.82	0.80	0.79	0.77	0.74	0.71	0.67	0.61	0.53	0.39	0.31	0.23	0.15	0.07	0.39
3	0.84	0.84	0.84	0.83	0.82	0.81	0.80	0.78	0.76	0.73	0.70	0.65	0.60	0.55	0.25	0.20	0.15	0.10	0.05	0.41
2	0.84	0.84	0.83	0.83	0.82	0.81	0.79	0.78	0.75	0.72	0.69	0.65	0.60	0.55	0.12	0.10	0.08	0.05	0.03	0.42
j=1	0.84	0.84	0.83	0.83	0.82	0.81	0.79	0.78	0.75	0.72	0.69	0.65	0.60	0.55	0.51	0.47	0.44	0.43	0.42	0.42

Figure 7.9. Non-dimensional potential function values.

e. Find pressures at point for each vertical distance from soil on and back of sheet pile.

We will use the dimensionless velocity potential values calculated above to determine the pressures on the sheet pile.

Water level difference, $\Delta H = 4.50$ m, maximum potential value, $\phi_{max} = 2.50$, water density, $\gamma = 9.81$ kN/m^3 are the given information to solve the problem. The following formula can be used to determine the pressure at various points on a sheet pile.

$$p = \gamma(S\text{-}JL)$$

where p is the water pressure,

γ is the water density,

S is the static head,

J is the slope of head losses: $J = \Delta H/\phi_{max} = 4.50/2.5 = 1.80$

L is the difference of velocity potential: $L = \phi_{max} - \phi$.

Φ is the velocity potential: $\phi = f\phi_{max}$.

f is the non-dimensional velocity potential as can be seen in Figure 7.9. For each point on sheet pile, the pressure values are calculated and included in Table 7.1.

Table 7.1. Calculating seepage pressure at various points on a sheet pile

Point	Coordinates, m	Grid No	Static head, m	Slope of head losses	Velocity potential	Difference of velocity potential	Head losses, m	Pressure Head, m	Pressure, kN/m^2
	x, y	i, j	S	$J = \Delta H/\varphi_{ma}$	$\Phi = f\varphi_{max}$	$\varphi_{max} - \varphi = L$	JL	h	$P = \gamma h$
O	6.00, 11.00	13,23	0.00	-	-	-	-	0.00	0.00
B	6.00, 8.50	13, 18	2.50	1.80	2.50x1.00 = 2.50	0.00	0.00	2.50	24.53
C	6.00, 2.50	13, 6	8.50	1.80	2.50x0.74 = 1.85	0.65	1.17	7.33	71.91
D	6.50, 2.50	14, 6	8.50	1.80	2.50x0.36 = 0.90	1.60	2.88	5.62	55.13
E	6.50, 6.00	14, 14	4.50	1.80	2.50x0.00 = 0.00	2.50	4.50	0.00	0.00

Calculated pressure heads are shown in Figure 7.10.

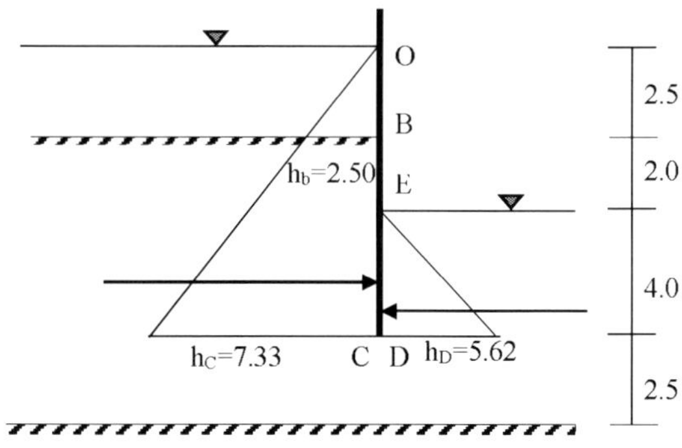

Figure 7.10. Pressure heads on the front and back of sheet pile.

7.5.3. Numerical Solution for Deflection of a Thin Plate

A Rectangular thin plate simply supported on all its four sides is subjected to a uniform lateral load triangular continuous load q. Using finite difference method, find the values of deflection at nodal points of the adopted network for the following given data.

Concrete plate dimensions: a = b = 6 m,
Uniform load: q = 4 kN/m^2
Thickness of the plate: t = 0.1m;
Concrete Poisson ratio: ν = 0.18;
Concrete Elasticity module: E = 2x10^7 kN/m^2.

The governing equation as an example of the Poisson equation:

$$\nabla^4 w = \frac{\partial^4 w}{\partial x^4} + 2\frac{\partial^4 w}{\partial x^2\, \partial y^2} + \frac{\partial^4 w}{\partial y^4} = \frac{P(x,y)}{D}$$

where, w is the small deflection,
P is the applied transverse load,
D is the flexural rigidity of the plate = $D = \frac{Et^3}{12(1-v^2)}$

Solution: For simply supported plate, the deflection is zero (w =0) on boundary condition at edges. Let us consider a thin plate in the form of prism with thickness t and two other dimensions a = b = 6 m as shown in Figure 7.11. In this plate with an elastic body, the theory of elasticity will be accepted to be valid. The plate thickness is less than one-tenth of other dimensions.

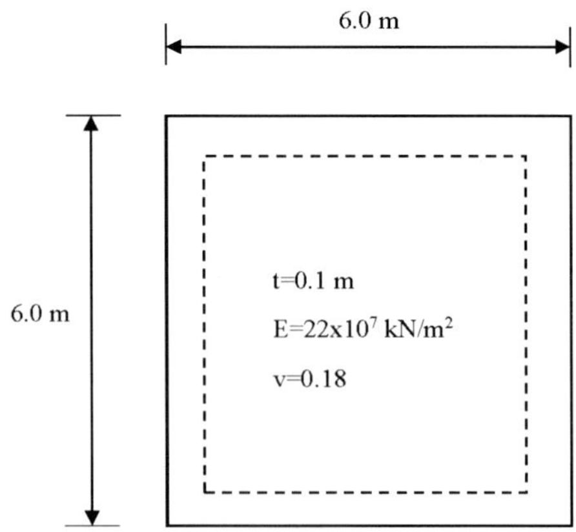

Figure 7.11. Two-way slab with simply supported edge.

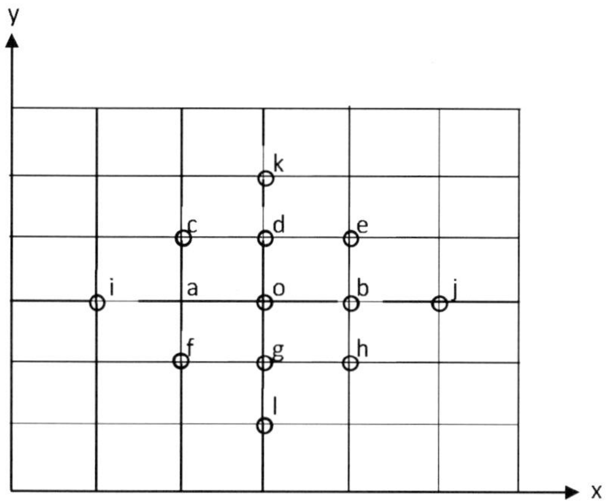

Figure 7.12. Mesh system for deflection calculation.

As shown in Figure 7. 20 we can choose $\Delta x = h$, $\Delta y = k$ as $h = k = 1.0$. Calculation of D for $E = 2x10^7$ kN/m^2,

$$D = \frac{Et^3}{12(1-\nu^2)} = 20000000 \times 0.1 \times 0.1 \times 0.1 \frac{1}{12(1-0.18\times 0.18)} = 1767.9$$

$$P(x, y)/D = (qh^4)/D = (4\times 1^4)/1767.9 = 0.002262624$$

We can obtain the following equilibrium equations by placing the link between stress and moment, stress and strain. According to the scheme in Figure 7.12, we can formulate difference Equation 7.6 in the finite difference form as:

$$\nabla^4 w = \frac{\partial^4 w}{\partial x^4} + 2\frac{\partial^4 w}{\partial x^2\, \partial y^2} + \frac{\partial^4 w}{\partial y^4} = 20w_0 - 8(w_a + w_b + w_d + w_g) + 2(w_c + w_e + w_f + w_h) + 1.0(w_i + w_j + w_k + w_l) = (q\times h^4)/D$$

or

$$20w_0 - 8(w_a + w_b + w_d + w_g) + 2(w_c + w_e + w_f + w_h) + 1.0(w_i + w_j + w_k + w_l) = 0.002262624$$

Computational molecules and their coefficients for deflection equations are shown in Figure 7.13.

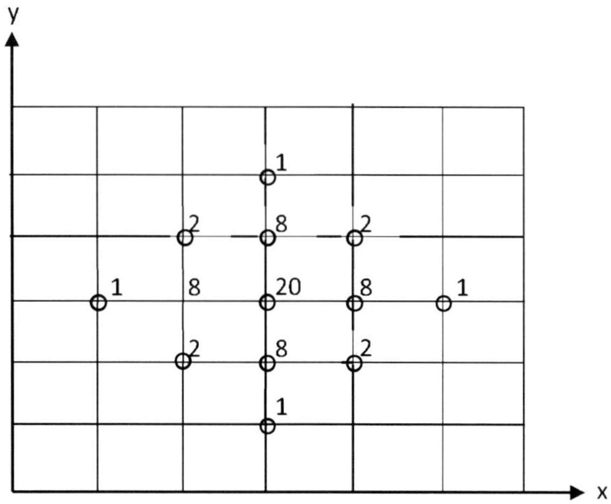

Figure 7.13. Computational molecule when h = k for deflection equation.

8	0	0	0	0	0	0	0	0	0
7	0	0	0	0	0	0	0	0	0
6	0	0	0	0	0	0	0	0	0
5	0	0	0	0	0	0	0	0	0
4	0	0	0	0	0	0	0	0	0
3	0	0	0	0	0	0	0	0	0
2	0	0	0	0	0	0	0	0	0
j =1	0	0	0	0	0	0	0	0	0
	0	0	0	0	0	0	0	0	0
		i = 1	**2**	**3**	**4**	**5**	**6**	7	**8**

Figure 7.14. Boundary condition values and first estimation values for all interior points.

In order to solve the deflection equation, we need some hypothetical boundary condition values at outside of the real boundaries. So, we have to use fictitious boundaries at out of plate edges. Boundary condition values and first estimation values for inner points are shown in Figure 7.14.

8			0.0030	0.0051	0.0058	0.0051	0.0031		
7		0.0000	0.0000	0.0000	0.0000	0.0000	0.0000	0.0000	
6	0.0030	0.0000	-0.0031	-0.0051	-0.0058	-0.0052	-0.0031	0.0000	0.0031
5	0.0051	0.0000	-0.0051	-0.0086	-0.0098	-0.0086	-0.0052	0.0000	0.0052
4	0.0058	0.0000	-0.0058	-0.0098	-0.0112	-0.0099	-0.0059	0.0000	0.0059
3	0.0051	0.0000	-0.0051	-0.0086	-0.0098	-0.0087	-0.0052	0.0000	0.0052
2	0.0031	0.0000	-0.0031	-0.0051	-0.0059	-0.0052	-0.0031	0.0000	0.0031
j = 1		0.0000	0.0000	0.0000	0.0000	0.0000	0.0000	0.0000	
			0.0031	0.0051	0.0059	0.0052	0.0031		
		i = 1	2	3	4	5	6	7	8

Figure 7.15. Deflection values computed after 800 iterations.

Computed deflection values after 800 iterations are shown in Figure 7.15.

As can be seen from the figure, we computed the maximum deflection for 800 iterations, at the midspan of the plate as 0.0112 m. We can compare this result with another method.

Using the classical solution from (Timoshenko and Woinowsky-Krieger, 1987).

Maximum deflection,

$$w_{max} = (0.00406\ qa^4)/D = (0.00406x4x6x6x6x6)/1767.9 = 0.011905\ \text{m}.$$

We see that there is a percentage difference of (0.011905 - 0.0112)/0.011905 = 0.059. There is 5.9 % difference between the answers from classical method and finite difference method.

7.5.4. Numerical Solution for Flow under a Dam

A dam with a base width of 30 m, is built on a permeable ground layer which has a 60 m thickness. Maximum water depth behind dam is equal to 10 m. Determine the boundary conditions for velocity potential for the problem and determine the velocity potential values (Figure 7.16).

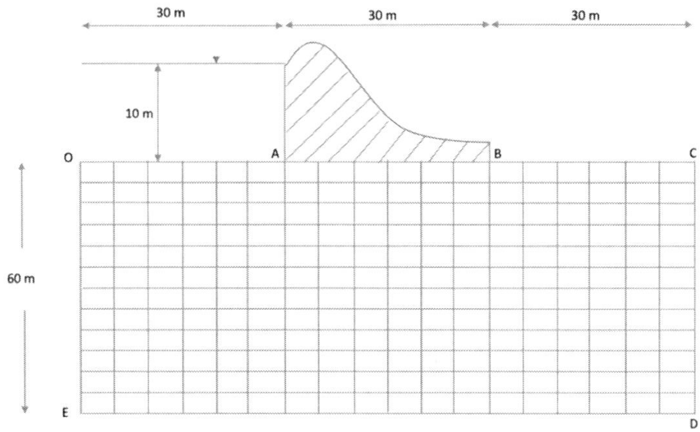

Figure 7.16. Computational grid to calculate the seepage flow under a dam.

Solution: To find the values of velocity potential function (φ) at inner points, the Laplace Equation is used as a governing equation. Computations are repeated for 500 iterations using Excel. Initial values are given as 5.0 for each nodes. The square mesh, $\Delta x = \Delta y = 5$ m, is used.

Boundary conditions: As can be seen from Figure 7.16, OA is the upstream part of the flow domain. The potential function on this boundary is constant and equals to water head above this line. Therefore, the boundary condition on $\overline{OA}$:

$$\varphi = 10.0$$

AB is a solid wall and is also a streamline. Therefore, the boundary condition for potential function on this line can be written on $\overline{AB}$:

$$\frac{\partial \varphi}{\partial y} = 0.0$$

As shown in Figure 7.16, BC is the downstream part of the flow domain. The potential function on this boundary is constant and equals to water head above this line. Therefore, the boundary condition on $\overline{BC}$:

$$\varphi = 0.0$$

ED is a solid wall and is also a streamline. Therefore, the boundary condition on this line can be written on $\overline{ED}$:

$$\frac{\partial \varphi}{\partial y} = 0.0$$

OE is the upstream boundary. We can assume a linear distribution boundary condition for potential function on $\overline{OE}$:

$$\frac{\partial \varphi}{\partial x} = 0.0$$

ED is the downstream boundary. We can also assume a linear distribution boundary condition for potential function on $\overline{ED}$:

$$\frac{\partial \varphi}{\partial x} = 0.0$$

By choosing 30 m on the upstream side of the dam body and 30 m on the downstream side, the total length of the computational area will be 90 m. The larger the computational area is selected, the more realistic the calculations around the dam will be. Since the permeable layer thickness is 60 m, calculations will be made in a 90mx60m computational area. A square mesh was selected and the grid spacing was chosen as 5 m. The potential function values calculated after 800 iterations are shown in Table 7.2. Since the calculation area is symmetrical with respect to the midpoint of the dam base, the calculated potential function values are also symmetrical. The equipotential lines of the flow can be obtained as in Figure 7.17.

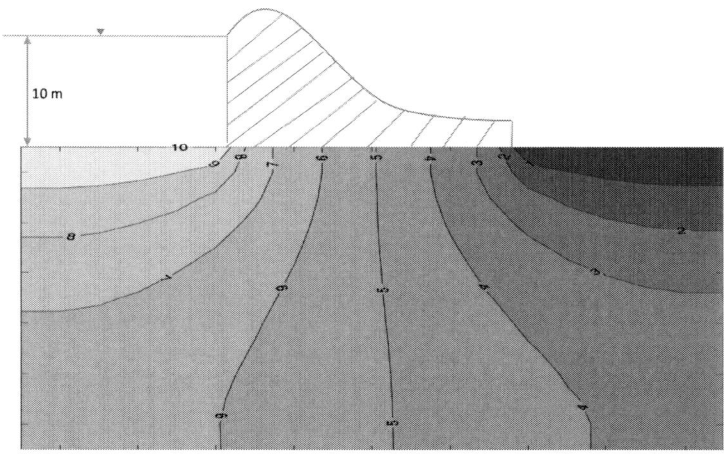

Figure 7.17. The graph of equipotential lines of the problem.

Table 7.2. The calculated equipotential lines

	i – 1	2	3	4	5	6	7	8	9	10	11	12	13	14	15	16	17	18	19
J – 1	10	10	10	10	10	10	7.38	6.51	5.76	5.05	4.35	3.59	2.7	0	0	0	0	0	0
2	9.36	9.36	9.32	9.22	9.01	8.55	7.38	6.51	5.76	5.05	4.35	3.59	2.69	1.5	1.03	0.81	0.71	0.67	0.67
3	8.76	8.76	8.69	8.54	8.27	7.81	7.09	6.38	5.71	5.06	4.4	3.72	3	2.27	1.79	1.51	1.36	1.29	1.29
4	8.23	8.23	8.15	7.99	7.72	7.32	6.79	6.22	5.64	5.06	4.48	3.89	3.31	2.78	2.37	2.09	1.92	1.84	1.84
5	7.78	7.78	7.7	7.54	7.3	6.95	6.53	6.06	5.57	5.06	4.56	4.06	3.59	3.16	2.81	2.56	2.4	2.32	2.32
6	7.4	7.4	7.33	7.18	6.96	6.67	6.32	5.92	5.5	5.07	4.63	4.21	3.81	3.45	3.16	2.94	2.79	2.72	2.72
7	7.09	7.09	7.03	6.9	6.71	6.45	6.15	5.81	5.45	5.07	4.7	4.33	3.99	3.68	3.43	3.24	3.1	3.04	3.04
8	6.85	6.85	6.79	6.68	6.5	6.28	6.02	5.72	5.41	5.08	4.75	4.43	4.13	3.87	3.64	3.47	3.35	3.3	3.3
9	6.66	6.66	6.61	6.51	6.35	6.16	5.92	5.66	5.37	5.08	4.79	4.5	4.24	4	3.8	3.65	3.55	3.49	3.49
10	6.53	6.53	6.48	6.38	6.24	6.06	5.85	5.61	5.35	5.08	4.82	4.56	4.32	4.1	3.92	3.78	3.68	3.63	3.63
11	6.44	6.44	6.39	6.3	6.17	6	5.8	5.58	5.34	5.09	4.83	4.59	4.37	4.17	4	3.86	3.77	3.73	3.73
12	6.4	6.4	6.35	6.27	6.14	5.97	5.78	5.56	5.33	5.09	4.84	4.61	4.39	4.2	4.03	3.91	3.82	3.77	3.77
13	6.4	6.4	6.35	6.27	6.14	5.97	5.78	5.56	5.33	5.09	4.84	4.61	4.39	4.2	4.03	3.91	3.82	3.77	3.77

7.5.5. Calculating Stream Function for a Potential Flow

Calculate the stream function values for a simple planar potential flow at a sharp-edged elbow as given in Figure 7.18. The width of the entrance is B = 5 m and the exit width A=8 m. External dimensions of the elbow is 14*14m. Accordingly, C = 6 m and E = 9 m.

Solution: Since the flow type is potential flow, the potential function must satisfy the Laplace equation as shown in Equation (7.3). Let's choose a square mesh of the form $\Delta x = \Delta y = h = 1$ m in the solution area.

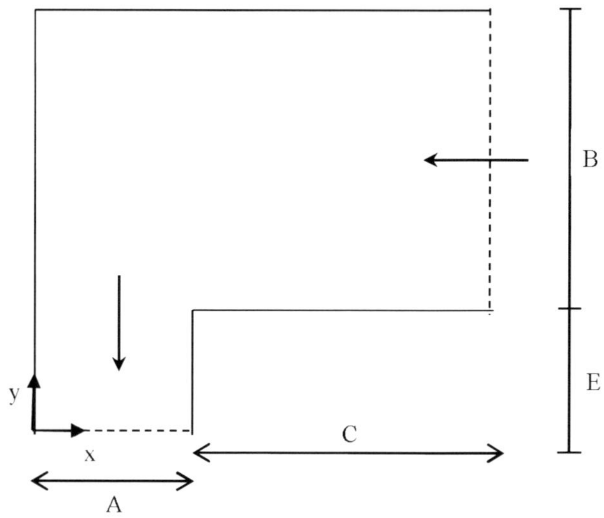

Figure 7.18. The flow in a sharp edged elbow.

Relation: The finite difference equivalent of Laplace's equation for a square mesh, as in Equation (7.12), can be calculated as:

$$\Psi_{i,j} = (\Psi_{i+1,j} + \Psi_{i-1,j} + \Psi_{i,j+1} + \Psi_{i,j-1}) / 4$$

Here, stream function, Ψ, at (i, j) node will be calculated by trial and error from the average of 4 neighborhood nodes. In the upper part of the flow domain, there are nodes from i=1 to 15, j=1 to 15, j=1 to 6, that is, a total of 15x6 = 90 nodes. In the lower part, there are i =1 to 9, j =7 to 15, i.e., 9x9 = 81 nodes. The total number of nodes is 171. Since the total number of nodes in solid boundaries is 45, 171- 45 = 126 is the number of interior points. Thus, 126 equations valid at 126 points in the computation domain will be solved. Since the scheme is implicit, the calculation will be made by iteration.

Boundary conditions: In a roughly 15x15 flow field, the boundary conditions are defined as:

On the outer limit of the elbow: Ψ = 1.00 (Dimensionless stream function).

On the inner border of the elbow: Ψ = 0.00.

Upstream boundary condition: Uniform velocity distribution is accepted. Accordingly, $\Psi_{i,j} = 2^* \Psi_{i-1,j} - \Psi_{i-2,j}$ can be written.

Downstream boundary condition: Uniform velocity distribution is accepted. Here $\Psi_{i,j} = 2^* \Psi_{i,j-1} - \Psi_{i,j-2}$ can be written. The initial estimate value for the interior nodes was taken as 0.5.

Since the calculation will be made with dimensionless stream function values, the boundary conditions will be determined accordingly. Solid walls (since the velocity components of the flow on the walls are 0.0, the dimensionless current function will be taken as constant in solid walls. The dimensionless stream function in a flow ranges from 0.0 to 1.0. If the dimensionless stream function is taken as Ψ=0.00 boundary condition at the lower limit of the current (inner limit of the elbow), the stream function value at the upper boundary (at the outer boundary of the elbow) will be Ψ = 1.00.

Table 7.3. Computed dimensionless stream function values after 100 iterations

	i = 1	2	3	4	5	6	7	8	9	10	11	12	13	14	15
j = 1	1	1	1	1	1	1	1	1	1	1	1	1	1	1	1
2	1	0.98	0.96	0.94	0.92	0.90	0.88	0.86	0.84	0.82	0.81	0.81	0.80	0.80	0.80
3	1	0.96	0.93	0.89	0.85	0.81	0.76	0.71	0.67	0.64	0.62	0.61	0.61	0.60	0.59
4	1	0.95	0.89	0.84	0.78	0.71	0.64	0.56	0.49	0.45	0.43	0.41	0.41	0.40	0.39
5	1	0.93	0.86	0.79	0.71	0.62	0.52	0.41	0.28	0.24	0.22	0.21	0.20	0.20	0.20
6	1	0.92	0.84	0.75	0.65	0.55	0.42	0.25	0.00	0.00	0.00	0.00	0.00	0.00	0.00
7	1	0.91	0.81	0.71	0.61	0.49	0.35	0.19	0.00						
8	1	0.90	0.80	0.69	0.57	0.45	0.32	0.16	0.00						
9	1	0.89	0.78	0.67	0.55	0.43	0.29	0.15	0.00						
10	1	0.89	0.77	0.65	0.53	0.41	0.28	0.14	0.00						
11	1	0.88	0.76	0.64	0.52	0.40	0.27	0.13	0.00						
12	1	0.88	0.76	0.64	0.51	0.39	0.26	0.13	0.00						
13	1	0.88	0.75	0.63	0.51	0.38	0.25	0.13	0.00						
14	1	0.87	0.75	0.62	0.50	0.37	0.25	0.12	0.00						
15	1	0.87	0.75	0.62	0.49	0.37	0.25	0.12	0.00						

It is assumed that the flow velocity is uniformly distributed in the inlet and outlet sections. After 100 iterations, dimensionless stream function values were calculated at all interior nodes and are shown in Table 7.3. Computed streamlines can be seen in Figure 7.19. Since the stream function is constant at the upper and lower limits of the flow, this upper limit and the lower limit are considered as streamlines.

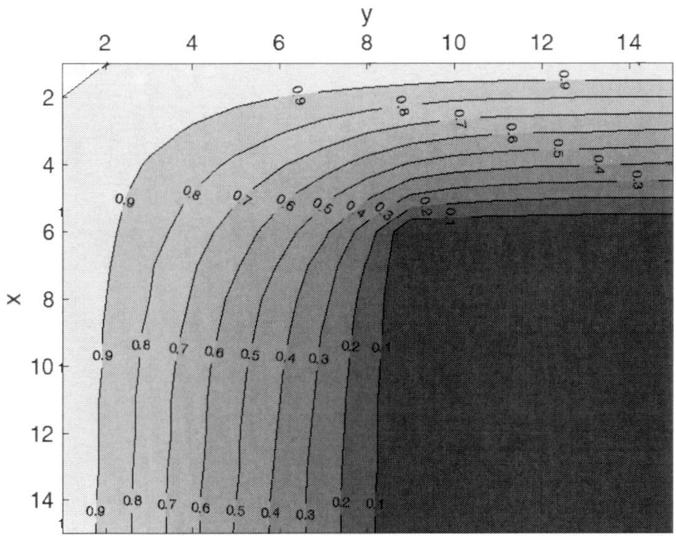

Figure 7.19. Computed streamlines for sharp edged elbow.

7.5.6. Calculation of Velocity Components of Flow from Stream Function Values in a Potential Flow

Dimensionless stream function values for a sharp corner are calculated in subsection 7.6.4 and shown in Table 7.3. The flowrate is 15 m^3/s and the flow width in the direction perpendicular to the flow is 2 m. Velocities at those points will be calculated from the given dimensionless stream function values.

a. Calculate the velocities at the interior nodes in the x-direction in the velocity field.
b. Calculate the velocities at the interior nodes in the y-direction in the velocity field.

c. Determine the resultant velocities and directions at the interior nodes.

Solution: Here are some explanations to understand the problem. It is possible to calculate other variables from known variables in the flow field.

1. If the velocity is known, the stream function can be determined.
2. If the velocity is known, the flowrate can be found.
3. If the velocity is known, the velocity potential function can be determined.
4. If the velocity is known, the vorticity can be found.
5. If the velocity is known, the water depth can be calculated.
6. If the velocity is known, the pressure can be found.

Similarly, if the stream function is known, other variables such as velocities can be found.

If the velocity components are $u = V_x$ and $v = V_y$, as can be seen in Figure 7.20 the resultant velocity, V is calculated by the following equation

$$V = \sqrt{V_x^2 + V_y^2} = \sqrt{u^2 + v^2}$$

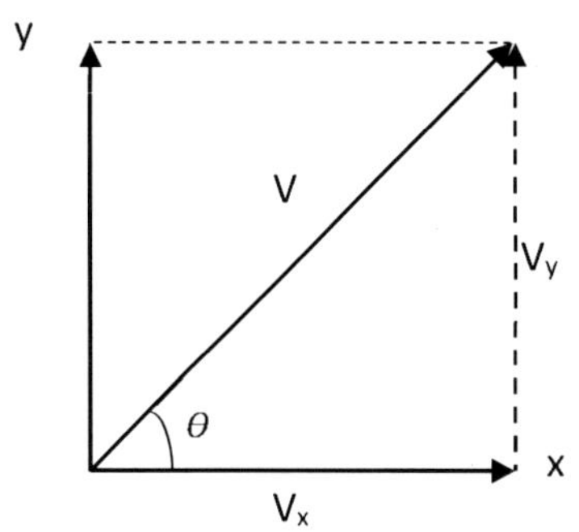

Figure 7.20. Resultant velocity.

$$\tan\theta = V_y/V_x = v/u$$

$$\theta = \tan^{-1}(v/u)$$

The stream function, which satisfies the continuity equation, satisfies the Laplace equation for irrotational flows. If the stream function is denoted by ψ (x, y), the velocities u and v in the x and y directions, respectively, are defined as:

$$u = \frac{\partial \psi}{\partial y}$$

$$v = -\frac{\partial \psi}{\partial x}$$

In this case, the defined stream function satisfies the continuity equation as shown below:

$$\partial u/\partial x + \partial v/\partial y = \partial^2\psi/\partial x\partial y - \partial^2\psi/\partial y\partial x = 0$$

So, any velocity field defined in this way represents the flow field of an incompressible fluid. Here, a combination of backward, forward, and central finite differences applied to the flow function ψ can be used to estimate the velocity components u and v. Where possible, derivatives should be estimated by central differences, but for nodes at the boundary it may be necessary to use finite forward or backward differences. Finite difference expressions for the velocity components defined above can be written as:

$$u_{i,j} = \frac{\psi_{i,j+1} - \psi_{i,j-1}}{2\Delta y}$$

$$v_{i,j} = -\frac{\psi_{i+1,j} - \psi_{i-1,j}}{2\Delta x}$$

The flow rate between points A and B can be written as:

$$Q = Dq = D\int_A^B v_n dn = D\int_A^B u dy - v dx = D\int_A^B \frac{\partial \psi}{\partial x} dx + \frac{\partial \psi}{\partial y} dy = D\int_A^B d\psi$$

$$Q/D = q = \psi_A - \psi_B$$

Here, q is the unit width flow rate (m^2/s), D is the flow width perpendicular to direction of flow (m), and v_n is the velocity component perpendicular to the flow direction (m/s). The last relation means that the difference of stream functions between points A and B is equal to the flowrate through the unit width through these points.

Velocities in the x-direction at the interior nodes in the flow field: The velocities in the x-direction calculated at the grid points are calculated and given in Table 7.4.

Table 7.4. Calculated velocity components, u, at x-direction

	i = 1	2	3	4	5	6	7	8	9	10	11	12	13	14	15
j = 1	0	0	0	0	0	0	0	0	0	0	0	0	0	0	0
2	0	0.13	0.27	0.41	0.57	0.73	0.91	1.08	1.24	1.34	1.41	1.45	1.48	1.5	1.52
3	0	0.13	0.26	0.4	0.55	0.72	0.91	1.11	1.3	1.4	1.45	1.48	1.49	1.5	1.51
4	0	0.12	0.24	0.37	0.52	0.69	0.89	1.15	1.45	1.51	1.52	1.51	1.51	1.5	1.49
5	0	0.11	0.22	0.33	0.46	0.62	0.82	1.15	1.84	1.68	1.6	1.55	1.52	1.5	1.48
6	0	0.09	0.19	0.28	0.38	0.49	0.62	0.79	0	0	0	0	0	0	0
7	0	0.08	0.15	0.22	0.29	0.36	0.39	0.34	0						
8	0	0.06	0.12	0.17	0.22	0.24	0.24	0.17	0						
9	0	0.05	0.09	0.13	0.15	0.16	0.15	0.09	0						
10	0	0.03	0.07	0.09	0.11	0.11	0.09	0.06	0						
11	0	0.03	0.05	0.07	0.08	0.08	0.06	0.04	0						
12	0	0.02	0.04	0.05	0.06	0.06	0.05	0.03	0						
13	0	0.02	0.03	0.04	0.05	0.05	0.04	0.02	0						
14	0	0.02	0.03	0.04	0.05	0.04	0.03	0.02	0						
15	0	0.02	0.02	0.04	0.05	0.04	0.03	0.02	0						

Similarly, the velocities in the y direction in the flow field were calculated and shown in Table 7.5.

Table 7.5. Calculated velocity components. v. at y-direction

	i = 1	2	3	4	5	6	7	8	9	11	12	13	14	15
j = 1	0.00	0.00	0.00	0.00	0.00	0.00	0.00	0.00	0.00	0.00	0.00	0.00	0.00	0.00
2	0.00	0.14	0.14	0.15	0.16	0.17	0.17	0.16	0.13	0.06	0.04	0.03	0.02	0.02
3	0.00	0.27	0.28	0.30	0.32	0.34	0.35	0.33	0.26	0.11	0.07	0.05	0.04	0.04
4	0.00	0.40	0.41	0.44	0.48	0.52	0.56	0.55	0.42	0.14	0.08	0.05	0.04	0.04
5	0.00	0.51	0.53	0.57	0.63	0.71	0.81	0.89	0.63	0.11	0.05	0.03	0.03	0.03
6	0.00	0.61	0.64	0.69	0.76	0.88	1.09	1.57	0.00	0.00	0.00	0.00	0.00	0.00
7	0.00	0.70	0.72	0.77	0.84	0.95	1.11	1.33	0.00					
8	0.00	0.77	0.79	0.83	0.89	0.97	1.08	1.18	0.00					
9	0.00	0.82	0.84	0.87	0.91	0.97	1.04	1.09	0.00					
10	0.00	0.85	0.87	0.89	0.93	0.97	1.01	1.04	0.00					
11	0.00	0.88	0.89	0.91	0.93	0.96	0.98	1.00	0.00					
12	0.00	0.90	0.91	0.92	0.94	0.95	0.96	0.97	0.00					
13	0.00	0.92	0.93	0.93	0.94	0.94	0.95	0.95	0.00					
14	0.00	0.94	0.94	0.94	0.94	0.94	0.94	0.94	0.00					
15	0.00	0.95	0.95	0.94	0.94	0.93	0.93	0.92	0.00					

Resultant velocities: The resultant velocities calculated from these found u and v velocities are given in Table 7.6 for each grid. In addition, the angle of each of the resultant velocities at the interior nodes is calculated and shown in Figure 7.21.

Table 7.6. Calculated resultant velocities

	i = 1	2	3	4	5	6	7	8	9	10	11	12	13	14	15
j = 1	0.00	0.00	0.00	0.00	0.00	0.00	0.00	0.00	0.00	0.00	0.00	0.00	0.00	0.00	0.00
2	0.00	0.54	0.60	0.69	0.80	0.93	1.07	1.22	1.33	1.41	1.45	1.48	1.50	1.52	1.52
3	0.00	0.75	0.79	0.87	0.97	1.09	1.23	1.37	1.49	1.52	1.53	1.52	1.52	1.53	1.51
4	0.00	0.90	0.94	1.01	1.11	1.23	1.38	1.56	1.72	1.67	1.61	1.56	1.54	1.53	1.49
5	0.00	1.02	1.06	1.12	1.21	1.34	1.52	1.76	2.16	1.82	1.66	1.58	1.54	1.52	1.48
6	0.00	1.11	1.15	1.20	1.29	1.42	1.61	1.94	0.00	0.00	0.00	0.00	0.00	0.00	0.00
7	0.00	1.18	1.21	1.26	1.33	1.42	1.54	1.67	0.00						
8	0.00	1.24	1.26	1.30	1.35	1.42	1.49	1.55	0.00						
9	0.00	1.28	1.30	1.32	1.36	1.40	1.45	1.48	0.00						
10	0.00	1.31	1.32	1.34	1.37	1.39	1.42	1.44	0.00						
11	0.00	1.33	1.34	1.35	1.37	1.39	1.40	1.41	0.00						
12	0.00	1.35	1.35	1.36	1.37	1.38	1.39	1.40	0.00						
13	0.00	1.36	1.36	1.36	1.37	1.37	1.38	1.38	0.00						
14	0.00	1.37	1.37	1.37	1.37	1.37	1.37	1.37	0.00						
15	0.00	1.38	1.38	1.37	1.37	1.37	1.36	1.36	0.00						

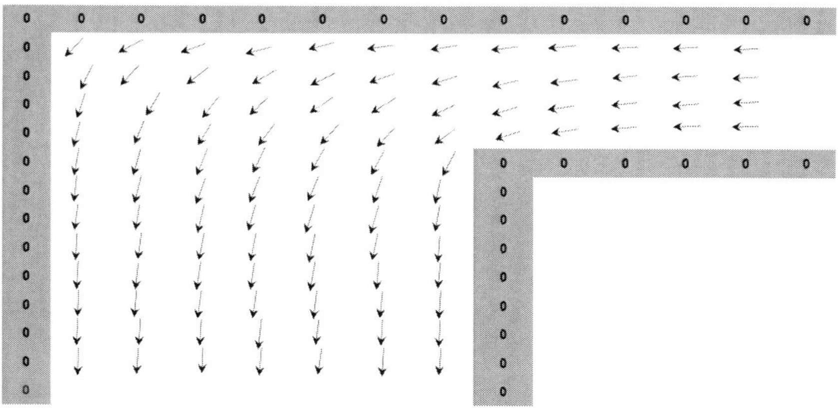

Figure 7.21. The velocity vectors in flow field.

Chapter 8

Solutions of Parabolic Equations

8.1. Introduction

Parabolic partial differential equations are used in many areas of engineering, such as civil, mechanical, environmental, and chemical engineering. The diffusion equation or the heat conduction equation are the typical parabolic differential equations used in hydraulics.

In this section, the solution of parabolic partial differential equations using the finite difference method is presented. The finite difference method can be easily used to solve problems in rectangular domains. However, it is more difficult to apply in irregularly shaped domains. We will see that there are some restrictions on the time interval in terms of numerical stability when explicit schemes are applied in finite differences (Causon and Mingham, 2010; Koutitas, 1983). Implicit schemes, however, are unconditionally stable but require longer computations.

8.2. Classification of Parabolic Partial Differential Equations

Parabolic equations can be classified in different ways.

1. Classification based on the order of partial differential equation:

The order of a partial differential equation is the highest-order partial derivative appearing in it. For instance, the advection equation is a first order partial differential equation.

$$\frac{\partial u}{\partial t} + \frac{\partial u}{\partial x} = 0 \tag{8.1}$$

On the other hand, the heat equation is considered a second order partial differential equation.

$$\frac{\partial u}{\partial t} + \frac{\partial^2 u}{\partial x^2} = 0 \tag{8.2}$$

2. Classification depends on the coefficient of the partial differential equation:

A partial differential equation is linear if the coefficients of the partial derivatives are not functions of u.

Let's consider the advection equation in Equation (8.1). Inserting a variable, u, as a multiplier to $\frac{\partial u}{\partial x}$, turns the advection equation into Burgers equation which is a nonlinear partial differential equation.

$$\frac{\partial u}{\partial t} + u\frac{\partial u}{\partial x} = 0 \tag{8.3}$$

3. Classification based on the value of discriminant of the second order partial differential equation:

The second order linear partial differential equation can be written in a general form as:

$$A\frac{\partial^2 u}{\partial x^2} + B\frac{\partial^2 u}{\partial x\, \partial y} + C\frac{\partial^2 u}{\partial y^2} + I(x, y, u, \frac{\partial u}{\partial x}, \frac{\partial u}{\partial t}) = 0 \tag{8.4}$$

These equations are classified based on the value of the discriminant: $B^2 - 4AC$.

Let us consider One Dimensional Advection-Diffusion Equation:

$$\frac{\partial u}{\partial t} = k\frac{\partial^2 u}{\partial x^2} + D(x, u, \frac{\partial u}{\partial t}) \tag{8.5}$$

In Equation (8.1) since the coefficients are as A = k, B = 0, C = 0, Equation (8.5) provides,

$$B^2 - 4AC = 0 \tag{8.6}$$

Thus, Equation (8.5) is a parabolic partial differential equation. Parabolic partial differential equations include only first derivative in time.

8.3. Schemes for Parabolic Partial Differential Equations

Some schemes are used to solve parabolic partial differential equations using the finite difference method. Many of these schemes are also used for numerical solution of hyperbolic partial differential equations which will be discussed in Chapter 9.

It is important to choose a suitable scheme to solve the problem. Selection of a suitable scheme depends on the governing equation of the problem, and its boundary and initial conditions. After selecting the finite difference scheme, the corresponding expression of finite difference scheme, which is equivalent to governing equation, could be solved (Yanovsky, 2005).

We can consider the simplest form of the parabolic equation as.

$$\frac{\partial u}{\partial t} = c\frac{\partial^2 u}{\partial x^2} \tag{8.7}$$

This is the homogeneous heat equation. However, for better understanding, we examine the inhomogeneous heat equation as a typical parabolic equation.

$$\frac{\partial u}{\partial t} = c\frac{\partial^2 u}{\partial x^2} + f \tag{8.8}$$

There are some alternative schemes to solve parabolic partial differential equation. In order to formulate the finite difference schemes we consider Equation (8.8).

8.3.1. Forward in Time and Centered Space (FTCS) Scheme

The forward in time and centered space scheme is also called the forward Euler scheme. The finite difference form of the heat equation can be obtained from Equation (8.8) using the forward in time and centered in space scheme as,

$$\frac{u_i^{n+1} - u_i^n}{\Delta t} = c\frac{u_{i+1}^n - 2u_i^n + u_{i-1}^n}{\Delta x^2} + f_i^n \tag{8.9}$$

The FTCS is conditionally stable and the order of accuracy is $O(\Delta t)+O(\Delta x^2)$ which is first order accurate in time and second order accurate in space. The general view of the grid system and used nodes in the solution of heat equation is given in Figure 8.1.

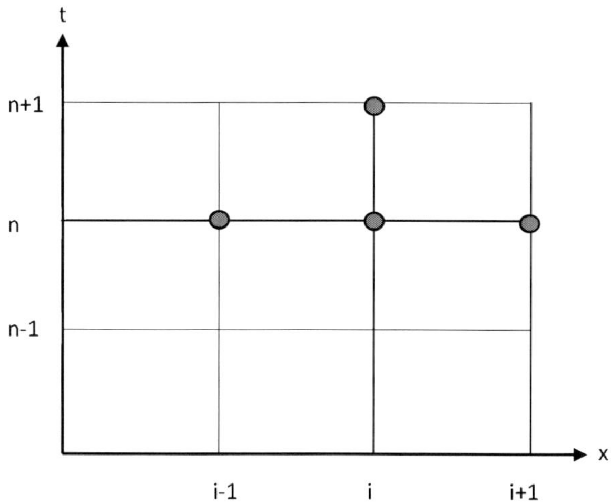

Figure 8.1. Grid points for FTCS (2 time level stencil).

8.3.2. Backward-Time Centered Space (BTCS) Scheme

The backward-time centered space scheme is also called the backward Euler scheme. We can write the expression of this scheme for point (i, n+1) as

$$\frac{u_i^{n+1}-u_i^n}{\Delta t}=c\,\frac{u_{i-1}^{n+1}-2u_i^{n+1}+u_{i-1}^{n+1}}{\Delta x^2}+f_i^n \tag{8.10}$$

This scheme is implicit and unconditionally stable. The order of accuracy is $O(\Delta t)+O(\Delta x^2)$.

8.3.3. Dufort-Frankel Scheme

The FTCS scheme is 1st order accurate in time and 2nd order accurate in space. To improve the scheme that is also 2nd order accurate in time, one can use central finite differences for $\partial u/\partial t$. The new scheme that is CTCS can be written as:

$$\frac{u_i^{n+1} - u_i^{n-1}}{2\Delta t} = c\frac{u_{i+1}^n + - 2u_i^n + u_{i-1}^n}{\Delta x^2} + f_i^n \tag{8.11}$$

This is an explicit equation and order of accuracy is $O(\Delta t^2) + O(\Delta x^2)$. However, the stability analysis shows that the CTCS schemes are unconditionally unstable. To make the CTCS scheme stable, the following expression is inserted to Equation (8.11).

$$u_i^n = \frac{u_i^{n+1} + u_i^{n-1}}{2} \tag{8.12}$$

So, the new scheme can be written as:

$$\frac{u_i^{n+1} - u_i^{n-1}}{2\Delta t} = c\frac{u_{i+1}^n - (u_i^{n+1} + u_i^{n-1}) + u_{i-1}^n}{\Delta x^2} + f_i^n \tag{8.13}$$

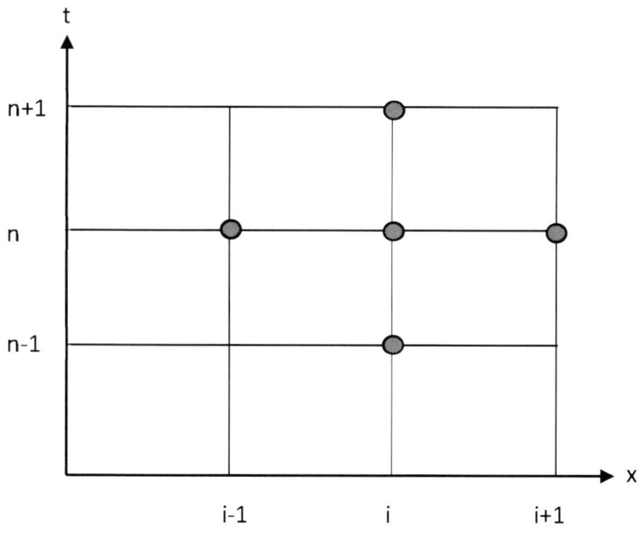

Figure 8.2. Grid points for Dufort-Frankel scheme (3 time level stencil).

This scheme is explicit and yet unconditionally stable. The order of accuracy is given by $O(\Delta x^2) + O(\Delta t^2) + O(\Delta x^2/\Delta t^2)$. The Dufort Frankel scheme is distinctive in that is as both explicit and unconditionally stable. However, there is a critical problem with this scheme. For the first time level, it is not possible to have previous time level value (Figure 8.2). It is appropriate to employ FTCS scheme for the first time step to achieve the solution, generally.

8.3.4. Crank-Nicolson Scheme

One of the disadvantages of the Dufort-Frankel scheme is that a special procedure should be applied at the initial time level, because the scheme includes three-time levels in the computations. Therefore, it is required to represent the second order approximation for $\partial u/\partial t$ with two-time levels using a half time interval.

$$\left.\frac{\partial u}{\partial t}\right|_i^{n+\frac{1}{2}} = \frac{u_i^{n+1} - u_i^n}{\Delta t} + O(\Delta t)^2 \quad (8.14)$$

Now, the problem turns out to approximate $\left.\frac{\partial u}{\partial t}\right|_i^{n+\frac{1}{2}}$ without including any term at half time $(n+\frac{1}{2})$ in the scheme. Also, the following expression is replaced in the space derivative.

$$u^{n+\frac{1}{2}} = \frac{u^n + u^{n+1}}{2} \quad (8.15)$$

which yields to:

$$\left.\frac{\partial^2 u}{\partial x^2}\right|_i^{n+\frac{1}{2}} = \frac{u_{i+1}^{n+\frac{1}{2}} + u_{i-1}^{n+\frac{1}{2}} - 2u_i^{n+\frac{1}{2}}}{\Delta x^2} + O(\Delta x)^2 \quad (8.16)$$

So, the new scheme is given by:

$$\frac{u_i^{n+1} - u_i^n}{\Delta t} = c\frac{1}{2}\frac{u_{i+1}^{n+1} - 2u_i^{n+1} + u_{i-1}^{n+1}}{\Delta x^2} + c\frac{1}{2}\frac{u_{i+1}^n - 2u_i^n + u_{i-1}^n}{\Delta x^2} + \frac{(f_i^{n+1} + f_i^n)}{2} \quad (8.17)$$

The Crank-Nicolson scheme is implicit and unconditionally stable. The order of accuracy is $O(\Delta t^2)+O(\Delta x^2)$. Figure 8.3 shows the grid points used in the Crank- Nicolson scheme.

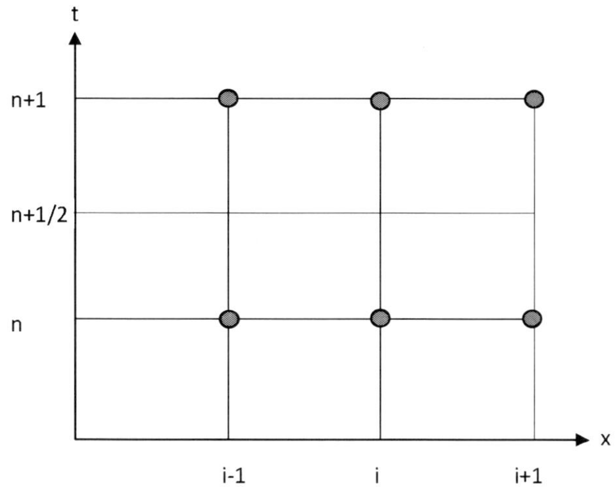

Figure 8.3. Grid points for the Crank-Nicolson scheme (2 time level stencil).

8.3.5. Leapfrog Scheme

The leapfrog scheme is a numerical integration method used to solve ordinary differential equations (ODEs) or partial differential equations (PDEs) that involve time evolution. It is a second-order accurate method, which means that the error in the solution grows quadratically with the time step, making it more accurate than first-order methods like the forward Euler method.

The basic idea behind the leapfrog scheme is to update the values of the solution at alternate time steps, so that the solution "leapfrogs" over the intermediate time points. The scheme uses a combination of half-step updates for both the time and the spatial variables, resulting in a more accurate and stable solution compared to some other numerical methods.

The leapfrog scheme is typically used for problems that have oscillatory behavior or for problems that involve wave propagation, as it has certain properties that make it well-suited for capturing such phenomena accurately. Here's how the leapfrog scheme works in its most basic form:

1. Start with the initial conditions for the variables of interest at time $t = 0$.

2. Compute the values of the variables at the first time step, t = dt/2, using a half-step update in time, where dt is the time step.
3. Compute the values of the variables at the next full time step, t = dt, using a full-step update in time.
4. Repeat step 2 and 3 for subsequent time steps until the desired time is reached.

$$\frac{u_i^{n+1} - u_i^{n-1}}{2\Delta t} = c\frac{u_{i+1}^n - 2u_i^n + u_{i-1}^n}{\Delta x^2} + f_i^n \qquad (8.18)$$

As seen from Figure 8.4 the CTCS scheme is unstable. To make CTCS scheme stable, one can employ leapfrog scheme.

The leapfrog scheme uses a centered difference approximation for the spatial derivatives, which means that it computes the derivatives at the midpoints between the grid points. This can result in more accurate solutions compared to forward or backward difference approximations, which only use values at the current time step or the previous time step, respectively.

One important consideration when using the leapfrog scheme is the stability condition, which dictates the maximum time step that can be used to ensure numerical stability. The stability condition depends on the specific problem being solved and the discretization used, and it must be satisfied to obtain accurate and reliable results.

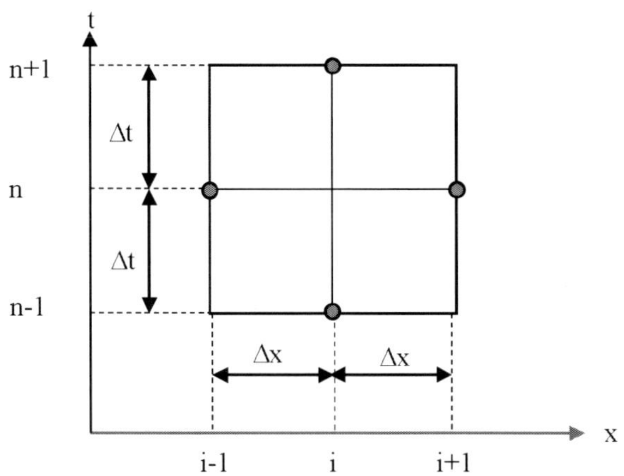

Figure 8.4. Grid points for the leapfrog scheme.

Overall, the leapfrog scheme is a popular numerical method for solving time-dependent problems, especially those involving wave propagation or oscillatory behavior, due to its accuracy and stability properties. However, it may not be suitable for all types of problems, and careful consideration of the specific problem and its requirements is necessary when choosing a numerical integration method.

The parabolic equation in two dimensions (x, y, t)

$$\frac{\partial u}{\partial t} = c\left(\frac{\partial^2 u}{\partial x^2} + \frac{\partial^2 u}{\partial y^2}\right) \tag{8.19}$$

is treated in the same way. The application of the explicit scheme presented permits the direct solution of the deriving uncoupled algebraic equations for the computation of u^{n+1} values from u^{n}.

The application of the Crank-Nicolson or any other implicit scheme, permitting the increase of the Δt step beyond the limit posed by stability criterion ($\lambda = v\Delta t/(\Delta z)^2$).

8.3.6. ADI Method

Using finite difference method, we obtain some algebraic equations from the governing equation. In these algebraic equations, the number of unknowns is equal to the number of the nodal points in the computation grid. For 2-D or 3-D solution domains this number becomes so large. Even if it is not prohibitive for contemporary computers, it makes the numerical solution computationally uneconomical. This problem is solved by introduction of the ADI method (Alternating Directions Implicit).

This technique, as its name suggests, consists of alternately applying implicit and explicit numerical schemes along the x and y dimensions.

This result in the need to solve, alternately, a n_{max} number of systems of equations in i_{max} unknowns at one time step and an i_{max} number of systems of equations in n_{max} unknowns at the next time step.

If the Crank-Nicolson scheme is used interchangeably with an explicit method, the ADI procedure can be described by the following equation.

$$\frac{u^{n+1} - u^{n}}{\Delta t} = c\frac{\partial^2 u^{n+1}}{\partial x^2} + c\frac{\partial^2 u^{n+\frac{1}{2}}}{\partial y^2} \tag{8.20}$$

and subsequently

$$\frac{u^{n+2} - u^{n+1}}{\Delta t} = c\frac{\partial^2 u^{n+1}}{\partial x^2} + c\frac{\partial^2 u^{n+\frac{3}{2}}}{\partial y^2} \quad (8.21)$$

where

$$u^{n+\frac{1}{2}} = \frac{u^n + u^{n+1}}{2} \quad (8.22)$$

$$u^{n+\frac{3}{2}} = \frac{u^{n+2} + u^{n+1}}{2} \quad (8.23)$$

8.4. Comparison of Schemes for Numerical Solution of Parabolic Equations

All schemes used in parabolic partial differential equations can be classified as

1. Explicit schemes,
2. Implicit schemes.

In general, explicit schemes are unstable, whereas implicit schemes are stable. Therefore, the explicit schemes should satisfy the stability condition, which are shown in Table 8.1. Order of accuracy is also shown in this table for each scheme. Comparison of schemes for numerical solution of parabolic equations is shown in Table 8.1.

Table 8.1. Comparison of schemes for numerical solution of parabolic equations

No	Schemes	Type	Stability	Order of accuracy
1	Forward Time – Centered Space	Explicit	Stable for $\lambda \leq 0,5$; $\lambda = c\frac{\Delta t}{\Delta x2}$	$O(\Delta t)$; $O(\Delta x)^2$
2	Backward Time-Centered Space	Implicit	Unconditionally stable	$O(\Delta t)$; $O(\Delta x)^2$
3	The Dufort-Frankel	Explicit	Unconditionally stable	$O(\Delta t)^2$; $O(\Delta x)^2$
4	The Crank-Nicolson	Implicit	Unconditionally stable	$O(\Delta t)^2$; $O(\Delta x)^2$
5	The Leap frog	Explicit	Unstable	$(\Delta t)^2$: 2; $(\Delta x)^2$: 2
6	ADI	Implicit and Explicit	Unconditionally stable	

8.5. Stability for Parabolic Partial Differential Equations

We consider the simplest form of the parabolic partial differential as can be seen in Equation (8.7), where, u is the rate of dissipation of the scalar magnitude, t is the time coordinate, x is the space coordinate c is the diffusion coefficient. The explicit method is conditionally stable for the 1-D spatial problem. The following equation can be written from Equation (8.7) as,

$$\frac{\partial T}{\partial t} = k\frac{\partial^2 T}{\partial x^2} \quad (8.24)$$

where T is the temperature, and k is the thermal conductivity.

The following is the stability condition for this problem:

$$\lambda = k\Delta t/(\Delta x)^2 \leq 1/2 \quad (8.25)$$

or stability criterion

$$\Delta t \leq (\Delta x)^2/2k \quad (8.26)$$

If $\lambda \leq 1/2$ can still yield oscillation (1D).
If $\lambda \leq 1/4$ ensures no oscillation (1D).
If $\lambda = 1/6$ tends to optimize truncation error.

For the 2-D spatial problem, for Equation (8.19), The following is the stability condition for explicit scheme:

$$\Delta t \leq [(\Delta x)^2 + (\Delta y)^2]/8k \quad (8.27)$$

If $h = \Delta x = \Delta y = h$;

$$\Delta t \leq h^2/4k \quad (8.28)$$

As a conclusion:

1. Solution of parabolic partial differential equations by explicit schemes:
    - It has very easy calculations.

- However, it has low accuracy, $O(\Delta t)$ accurate with respect to time, and subjects to instability.
- Therefore, It is necessary to use "small" Δt's. The iteration requires many steps.

2. The implicit solution of parabolic partial differential equations is unconditionally stable. As a result, Δx and Δt can be chosen only to control overall accuracy.
3. The implicit Crank-Nicolson approach offers 2nd-order accuracy in both space and time.
4. The ADI method offers a technique for solving parabolic equations in two spatial dimensions using tridiagonal matrices.

8.6. Applications of Parabolic Partial Differential Equations

8.6.1. Solution of Parabolic Partial Differential Equation Using Explicit Finite Differences

Using explicit finite difference method, consider heat conduction on a thin rod for the following data;

Given data: Heat Conductivity coefficient, $k = 0.90$ cm^2/s; Rod length: $L = 10$ cm.

Initial conditions: $T(0 < x < 10, t = 0) = 10°$

Boundary conditions: $T(x = 0, t) = 80°$; $T(x = 10, t) = 40°$.

a. Find the temperature distribution on the rod for $\Delta t = 1$ secs, $\Delta x = 2.0$ cm, by hand, for total time 5 s.
b. Find the temperature distribution on the rod for $\Delta t = 1$ secs, $\Delta x = 2.0$ cm, by computer for total time 5 s.
c. Find the temperature distribution on the rod for $\Delta t = 1$ secs, $\Delta x = 2.0$ cm, by computer for total time 80 s.
d. Find the temperature distribution on the rod for $\Delta t = 10$ secs, $\Delta x = 2.0$ cm, by computer for total time 80 s.

Solution: Considering the initial temperature distribution as well as boundary temperatures, we can write the function $T(x, t)$ for 1-Dimensional heat equation as in Equation (8.7).

1. Consider the domain (see Figure 8.5).
2. Consider the heat conduction equation, 1-D Heat Equation, as in Equation (8.24).
3. Write the derivatives in terms of finite difference. Express a future state, T_i^{n+1}, only in terms of the present state, T_i^n.

For forward finite difference in time,

$$\frac{\partial T}{\partial t} = \frac{T_i^{n+1} - T_i^n}{\Delta t} + O(\Delta t)$$

For centered finite difference in space,

$$\frac{\partial^2 T}{\partial x^2} = k\frac{T_{i+1}^n + T_{i-1}^n - 2T_i^n}{\Delta x^2} + O(\Delta x)^2 +$$

Solving for T_i^{n+1}, results in:

$$T_i^{n+1} = T_i^n + \lambda(T_{i-1}^n - 2T_i^n + T_{i+1}^n)$$

where $\lambda = k\Delta t/(\Delta x)^2$

$$T_i^{n+1} = (1 - 2\lambda)T_i^n + \lambda(T_{i-1}^n + T_{i+1}^n)$$

4. Discretize the domain into a grid of evenly spaces points (nodes). Select the space interval and time interval.

For $\Delta t = 1$ sec, $\Delta x = 2.0$ cm, $k = 0.90$ cm^2/s,

$\lambda = k\Delta t/(\Delta x)^2 = 0.225.$

5. Find the stability criterion. The explicit method is conditionally stable.

For the 1-D spatial problem, the following expression is the stability condition:

$\lambda = k\,\Delta t/(\Delta\,x)^2 \leq 1/2$

or stability criterion

$\Delta t \leq (\Delta x)^2 k/0.5$

$\Delta t \leq (2)^2 \times 0.9 \times 2$

$\Delta t \leq 7.2$ seconds.

Therefore, since the selected $\Delta t = 1$ s is less than 7.2, the solution is stable.

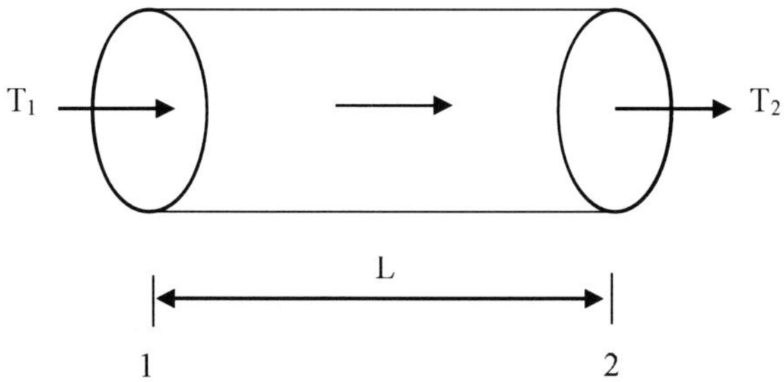

Figure 8.5. Definition of heat conduction on a thin rod.

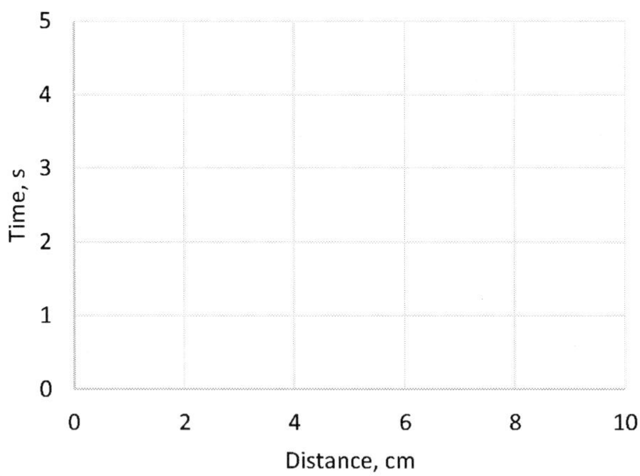

Figure 8.6. Selected coordinate system and computational grid.

Use the initial conditions and boundary conditions to solve the problem by systematically moving forward in time for $\Delta t = 1$ s. It is desired that express all future $(t + \Delta t)$ values, $T(x, t + \Delta t)$, in terms of current (t) and previous (t - Δt) information, which are known. In this problem we assumed that the origin of coordinate system starts as x = 0, and t = 0, as can be seen from Figure 8.6.

Initial and boundary conditions are shown in Figure 8.7.

80.0					**40.0**
80.0					**40.0**
80.0					**40.0**
80.0					**40.0**
80.0					**40.0**
80.0	**10.0**	**10.0**	**10.0**	**10.0**	**40.0**

Figure 8.7. Initial and boundary conditions.

The finite difference analog of the 1-D Heat Equation, Equation (8.24), can be expressed as

$$T_i^{n+1} = T_i^n + \lambda(T_{i-1}^n - T_i^n + T_{i+1}^n)$$

Calculation by hand:

Starting from t = 0 s (n = 0), find results at t = 1 s (n=1):

$$T_1^1 = T_1^0 + \lambda(T_0^0 - 2T_1^0 + T_2^0) = 10 + 0.225(80 - 2\times10 + 10) = 25.8°$$

$$T_2^1 = T_2^0 + \lambda(T_1^0 - 2T_2^0 + T_3^0) = 10 + 0.225(10 - 2\times10 + 10) = 10°$$

$$T_3^1 = T_3^0 + \lambda(T_2^0 - 2T_3^0 + T_4^0) = 10 + 0.225(10 - 2\times10 + 10) = 10°$$

$$T_4^1 = T_4^0 + \lambda(T_3^0 - 2T_4^0 + T_5^0) = 10 + 0.225(10 - 2\times10 + 40) = 16.8°$$

From t = 1 secs. (n = 1), find results at t = 2 secs. (n = 2):

$$T_1^2 = T_1^1 + \lambda(T_0^1 - 2T_1^1 + T_2^1) = 25.8 + 0.225(80 - 2\times25.8 + 10) = 34.4°$$

$$T_2^2 = T_2^1 + \lambda(T_1^1 - 2T_2^1 + T_3^1) = 10 + 0.225(25.8 - 2\times10 + 10) = 13.5°$$

$T_3^2 = T_3^1 + \lambda(T_2^1 - 2T_3^1 + T_4^1) = 10 + 0.225(10 - 2\times10 + 16.8) = 11.5°$

$T_4^2 = T_4^1 + \lambda(T_3^1 - 2T_4^1 + T_5^1) = 16.8 + 0.225(10 - 2\times(16.8) + 40) = 20.5°$

Similarly, we can calculate T values for t = 3, 4, and 5 secs. The calculated values of temperature by hand are shown in Table 8.2. The graph of the temperature distribution is shown in Figure 8.8.

Table 8.2. Calculated Temperature values for Δt = 1 s by hand

Distance, x, cm	Time			
	t = 0 s	t = 1 s	t = 3 s	t = 5 s
0	80	80	80	80
2	10	25.8	40	47.1
4	10	10	17.8	25.7
6	10	10	14	19.8
8	10	16.8	22.8	26.4
10	40	40	40	40

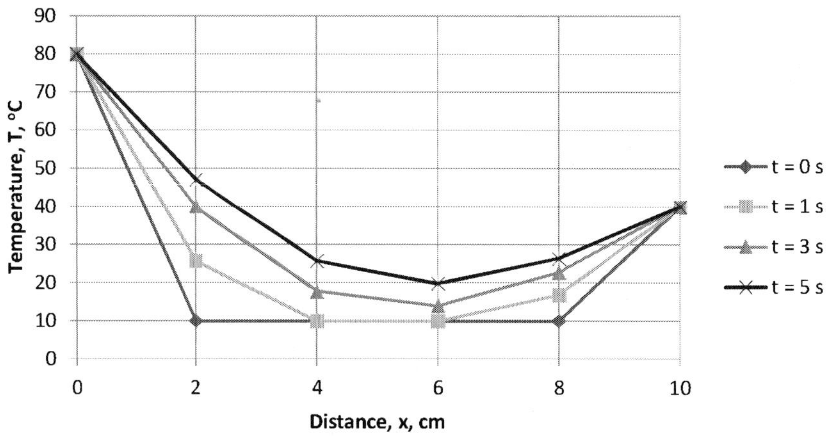

Figure 8.8. Temperature distribution calculated by hand for different time steps.

Instead of doing the calculations by hand, we can use a computer to compute the temperature values and manually verify the results.

b. Computer calculation for t = 1 s: The calculated values by the computer using an explicit scheme are identical to the determined values by hand in Table 8.2.

c. The heat distribution values are computed by computer and shown in Table 8.3. A graph for computed values is shown in Figure 8.9 for different durations.

Table 8.3. The heat distribution values

Distance, x, cm	Time					
	t = 0 s	t = 5 s	t = 15 s	t = 30 s	t = 45 s	t = 80 s
0	80	80	80	80	80	80
2	10	47.1	62.6	69.6	71.4	72
4	10	25.7	48.8	60	63	64
6	10	19.8	40.8	52	55	56
8	10	26.4	38.6	45.6	47.4	48
10	40	40	40	40	40	40

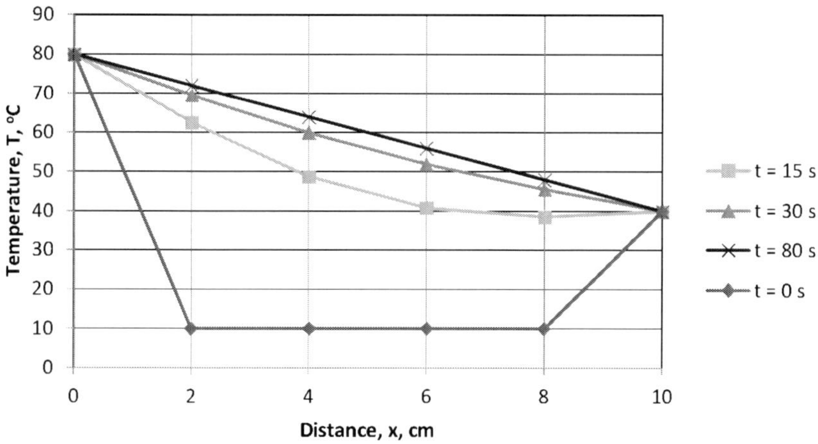

Figure 8.9. Heat distribution on the rod for Δt = 1 s using explicit scheme.

d. Solution for t =10 s (using explicit scheme by computer):

The calculated temperature values for (1, t) nodes are shown in Table 8.4. The graph of computed values is shown in Figure 8.10.

Table 8.4. Solution for Δt = 10 s by explicit scheme

Time, t, s	Calculated temperature, T, °C
0	10
10	168
20	-384
30	2343

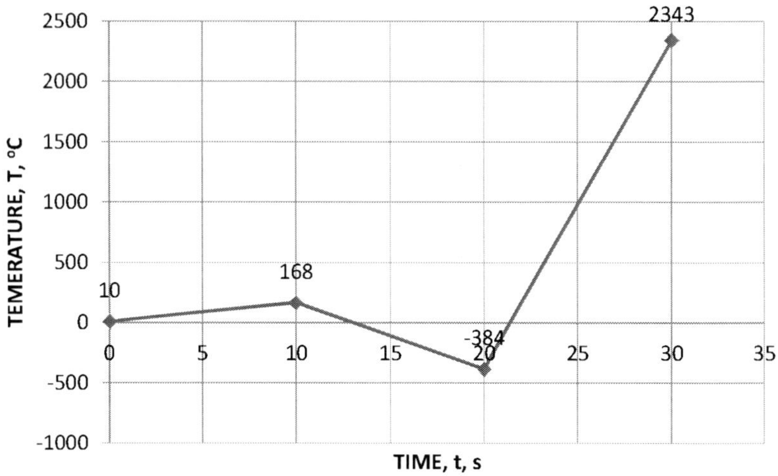

Figure 8.10. Computed temperature values for $\Delta t = 10$ secs using explicit scheme.

As can be seen from Table 8.4 and Figure 8.10, the solution is unstable for $\Delta t = 10$ s. Since the selected $\Delta t = 10$ s is greater than 7.2 s, the solution is unstable (see Figure 8.9). In fact, the temperature values should be between 40 and 80°C in all nodes.

8.6.2. Determination of the Stability Criterion for Flow Between Two Parallel Plates

The space between two parallel horizontal plate is filled with liquid that has the kinematic viscosity coefficient of $\nu = 1$ cm^2/s. The distance between two plates is 2 cm. The upper plate is linearly accelerated from a zero initial velocity to 30 cm/s in 0.1 s. The velocity profile variation of the liquid with time is to be computed.

Considering that there is no change in x- y plane, according to the notation shown in Figure 8.11, the variation of velocity profile, u(z, t), can be expressed by the following equation.

$$\frac{\partial u}{\partial t} = \nu \frac{\partial^2 u}{\partial z^2}$$

where U is the upper plate velocity. This is a parabolic equation similar to Equation (8.7). The flow domain is bounded between z = 0 cm and z = 2 cm.

The initial condition is $u(z,0) = 0$ cm/s. The boundary conditions are $u_{z=0} = 0$, $u_{z=2} = U$ cm/s,

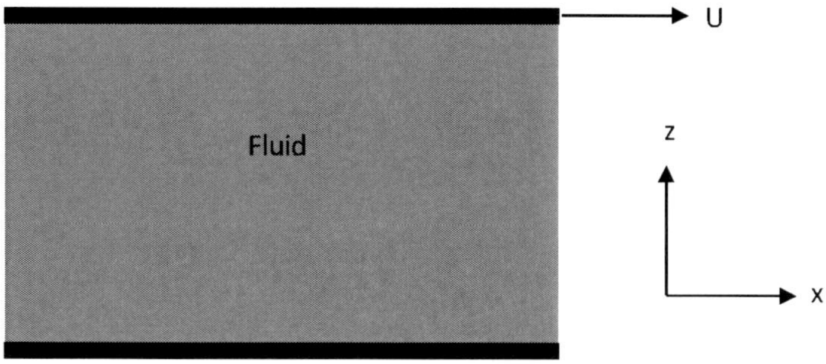

Figure 8.11. Description of the problem.

The numerical integration is performed by means of the explicit finite differences scheme described in above equation with $\Delta z = 0.2$ cm, and for two different time intervals $\Delta t = 0.01$ s and $\Delta t = 0.001$ s. The space discretization implies that the velocity is computed at 11 equidistant points between the plates along a straight line normal to the plates. The velocity is given by the boundary conditions i =1 and i = 11.

Stability criterion for the above parabolic equation can be written as:

$$\lambda = \nu \Delta t/(\Delta z^2)$$

$$\lambda \leq 1/2$$

a. For $\Delta z = 0.2$ cm, and $\Delta t = 0.01$ sec,

$$\lambda = 1 \times 0.01/0.2^2 = 0.25 \leq 0.5$$

This case yields a stable solution.

b. For $\Delta z = 0.2$ cm and $\Delta t = 0.1$ sec.

$$\lambda = 1 \times 0.1/0.04 = 2.5 > 0.5.$$

This case yields an unstable solution.

To find the velocity profiles for different time levels, the FTCS approximation is employed for case (a). For forward finite difference in time,

$$\frac{\partial u}{\partial t} = \frac{u_i^{n+1} - u_i^n}{\Delta t} + 0(\Delta t)$$

For centered finite difference in space,

$$\frac{\partial^2 u}{\partial x^2} = \nu \frac{u_{i+1}^n + u_{i-1}^n - 2u_i^n}{\Delta x^2} + O(\Delta x)^2 +$$

Solving for u_i^{n+1}, results in:

$$u_i^{n+1} = u_i^n + \lambda(u_{i-1}^n - 2u_i^n + u_{i+1}^n)$$

where $\lambda = \nu\ \Delta t/(\Delta x)^2$.

$$\lambda = \nu\ \Delta t/(\Delta x)^2 = 0.1x0.01/0.2^2 = 0.25$$

The upper plate velocity equation can be expressed with the following formula (as seen in Figure 8.12):

$U = 300t$, for $0 < t < 0.1$

$U = 30$, for $0.1 < t < \infty$.

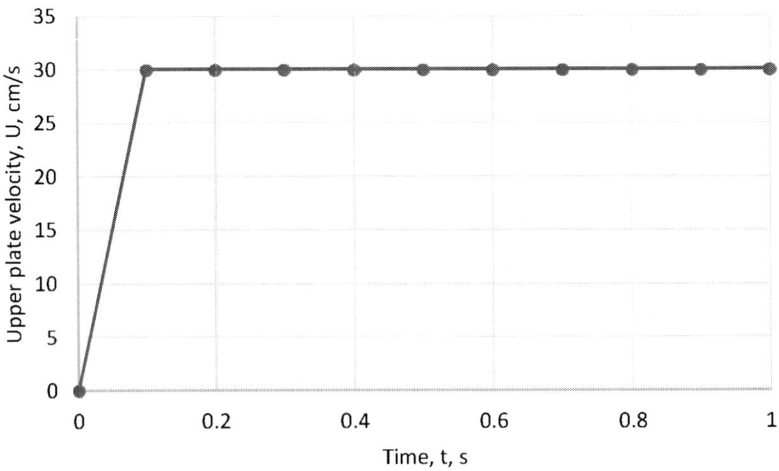

Figure 8.12. Variation of upper plate velocity with time.

So, the finite difference form of the problem is:

$$u_i^{n+1} = u_i^n + 0.25(u_{i-1}^n - 2u_i^n + u_{i+1}^n)$$

The velocity profiles are included in Table 8.5 up to 0.15s.

Table 8.5. Calculated velocity profiles, u(z, t) up to 0.15s

t (s)	z (cm)										
	0	0.2	0.4	0.6	0.8	1	1.2	1.4	1.6	1.8	2
0	0	0	0	0	0	0	0	0	0	0	0
0.01	0	0.00	0.00	0.00	0.00	0.00	0.00	0.00	0.00	0.00	3
0.02	0	0.00	0.00	0.00	0.00	0.00	0.00	0.00	0.00	0.75	6
0.03	0	0.00	0.00	0.00	0.00	0.00	0.00	0.00	0.19	1.88	9
0.04	0	0.00	0.00	0.00	0.00	0.00	0.00	0.05	0.56	3.23	12
0.05	0	0.00	0.00	0.00	0.00	0.00	0.01	0.16	1.10	4.76	15
0.06	0	0.00	0.00	0.00	0.00	0.00	0.05	0.36	1.78	6.40	18
0.07	0	0.00	0.00	0.00	0.00	0.01	0.11	0.64	2.58	8.15	21
0.08	0	0.00	0.00	0.00	0.00	0.04	0.22	0.99	3.49	9.97	24
0.09	0	0.00	0.00	0.00	0.01	0.07	0.37	1.42	4.48	11.86	27
0.10	0	0.00	0.00	0.00	0.02	0.13	0.56	1.92	5.56	13.80	30
0.11	0	0.00	0.00	0.01	0.05	0.21	0.79	2.49	6.71	15.79	30
0.12	0	0.00	0.00	0.02	0.08	0.32	1.07	3.12	7.93	17.07	30
0.13	0	0.00	0.01	0.03	0.12	0.44	1.40	3.81	9.01	18.02	30
0.14	0	0.00	0.01	0.05	0.18	0.60	1.76	4.51	9.96	18.76	30
0.15	0	0.00	0.02	0.07	0.25	0.79	2.16	5.18	10.80	19.37	30

The calculated velocity profiles for different time levels are included in Table 8.6 and also shown in Figure 8.13.

It is clear from the results that the velocity profiles become more linear as time increases. Theoretically, a linear velocity profile is obtained when time goes to infinity.

Table 8.6. Calculated velocity profiles for different time levels

t (s)	z (cm)										
	0	0.2	0.4	0.6	0.8	1	1.2	1.4	1.6	1.8	2
0.05	0	0.00	0.00	0.00	0.00	0.00	0.01	0.16	1.10	4.76	15
0.10	0	0.00	0.00	0.00	0.02	0.13	0.56	1.92	5.56	13.80	30
0.20	0	0.03	0.11	0.32	0.84	2.00	4.24	8.07	13.82	21.36	30
0.50	0	1.12	2.40	4.00	6.07	8.70	11.94	15.79	20.19	24.99	30
0.75	0	1.96	4.01	6.26	8.78	11.61	14.77	18.25	22.00	25.95	30
1.00	0	2.44	4.93	7.52	10.26	13.17	16.26	19.52	22.93	26.44	30

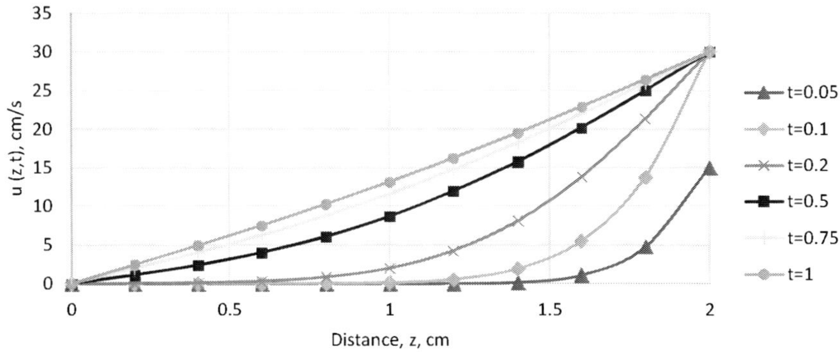

Figure 8.13. Calculated velocity profiles for different time level.

8.6.3. Numerical Solution of Heat Flow Equation

A large flat steel plate is 2 cm thick. The initial temperature within plate are given, as a function of distance from one face, by the following equations (Asmar, 2004):

$$U = 50x, \text{ for } 0 \leq x \leq 1$$

and

$$U = 50(2 - x), \text{ for } 1 \leq x \leq 2$$

Find the temperatures as a function of x and t if one face is maintained at 10^0C and the other face is 0^0C. The dimensional heat flow equation is given as follows.

$$\frac{\partial u}{\partial t} = k\frac{\partial^2 u}{\partial x^2}$$

Take $\Delta x = 0.20$ cm and heat conductivity coefficient, $k = 0.854$ cm^2/s.

Solution: The finite difference expression (FTCS) of the above equation can be written as

$$\frac{ui^{n+1} - u_i^n}{\Delta t} = \frac{u_{i+1}^n + u_{i-1}^n - 2u_i^n}{\Delta x^2}$$

$$u_i^{n+1} - u_i^n = k\frac{\Delta t}{\Delta x^2}(u_{i+1}^n + u_{i-1}^n - 2u_i^n)$$

$$u_i^{n+1} = u_i^n + k\frac{\Delta t}{\Delta x^2}(u_{i+1}^n + u_{i-1}^n - 2u_i^n)$$

Stability criterion:

$$\Delta t \leq \frac{1}{2}\frac{\Delta x^2}{\alpha}$$

$k = 0.854$

For $\Delta x = 0.2$,

$$k\frac{\Delta t}{\Delta x^2} < \frac{1}{2} \rightarrow \Delta t \leq \frac{1}{2}\frac{\Delta x^2}{k}$$

Stability criterion,

$$\Delta t \leq \frac{1}{2}\frac{(0.2)^2}{0.854} \rightarrow \Delta t \leq 0.023 \text{ s}$$

Let's select $\Delta t = 0.02\ s$.

Boundary and initial conditions are shown in Figure 8.14.

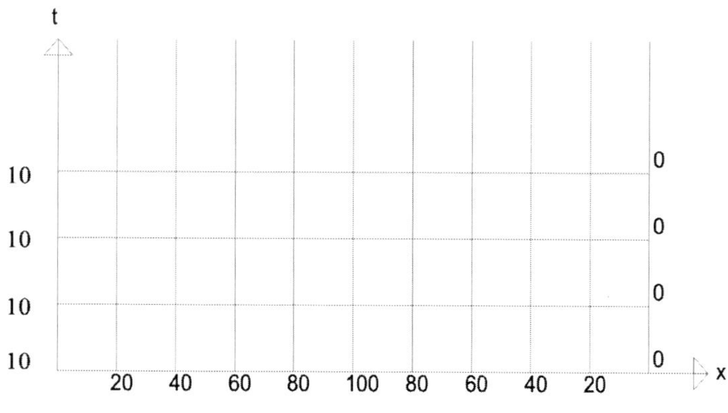

Figure 8.14. Initial and boundary conditions.

Finite difference equivalent can be written as:

$$u_i^{n+1} = u_i^n + \frac{0.02\times0.854}{(0.2)^2}(u_{i+1}^n + u_{i-1}^n - 2u_i^n)$$

Solving this equation yields

$$u_i^{n+1} = u_i^n + 0.43(u_{i+1}^n + u_{i-1}^n - 2u_i^n)$$

The resulting distribution of heat after 3 s is given in Figure 8.15

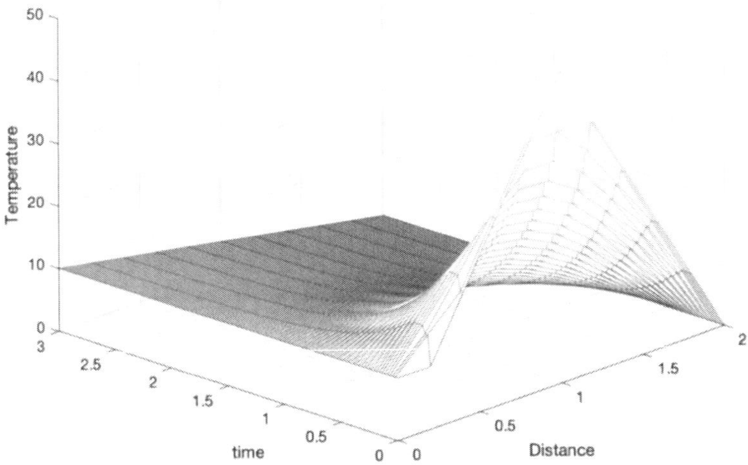

Figure 8.15. Distribution of heat between to faces of plate for t=3 s.

Chapter 9

Solutions of Hyperbolic Equations

9.1. Introduction

In this chapter, numerical solutions of hyperbolic partial differential equations encountered in engineering, especially hydraulic engineering will be studied. These equations are used to clarify many events occurring in vibrations such as in buildings, beams and machines, and wave theory, in industry, hydrodynamics, heat transfer, mass transfer. In fact, the hyperbolic partial differential equations govern all propagation problems. Unlike elliptic equations, which describe processes that are dominated by diffusion, hyperbolic equations describe processes that involve waves or waves of change.

The hyperbolic partial differential equations are the most difficult to solve among the differential equations. The solutions to hyperbolic equations often involve the use of special functions, such as the sinh and cosh functions, and they can be challenging to solve due to the complexity of the equations and the boundary conditions that must be satisfied. Despite these challenges, however, hyperbolic equations are an important tool for understanding and predicting the behavior of many physical systems (Gary, 1969). For this reason, hyperbolic partial differential equations are sometimes solved using the characteristics method. This subject will be examined in a separate section called the method of characteristics.

In the chapter, after a general classification, the properties of hyperbolic partial differential equations, the properties of advection-diffusion equation, finite difference schemes used in hyperbolic partial differential equations will be explained, and then some example problems will be solved numerically.

9.2. Hyperbolic Partial Differential Equations

Let's consider a directional wave equation that is an example of all hyperbolic partial differential equations.

$$\frac{\partial f}{\partial t} + c\frac{\partial f}{\partial x} = 0 \tag{9.1}$$

where: f is the function,

c is a constant,

t is the time variable,

x is the space variable.

This equation can be solved analytically depending on initial condition. For initial condition, the initial time is always taken to be 0. Therefore, f(0, x) is required to be equal to a given function $f_0(x)$ for all values of x. The values of f(t, x) can be determined for positive values of t. This problem is referred as an initial value problem.

The solution of (9.1) is given as,

$$f = f_o(x - ct) \tag{9.2}$$

This equation represents a wave propagating with a constant velocity c. The shape of the wave stays constant. Actually, it is a pure translation. If $c > 0$, the wave propagates in the positive x-direction, whereas if $c < 0$, the wave propagates in the negative x-direction.

Equation (9.2) may serve as a model-equation for a compressible fluid, namely, if *f* denote the water depth it represents a water wave propagating with the velocity c.

To allow for generalization, the equation (9.2) can be written in the following form:

$$\frac{\partial f}{\partial t} + \frac{\partial F}{\partial x} = 0 \tag{9.3}$$

where for the linear advection equation F(c) = cf.

Another hyperbolic equation, a good example to exhibit the properties of the hyperbolic type of equations, is the linear wave equation:

$$\frac{\partial^2 f}{\partial t^2} - c^2\frac{\partial^2 f}{\partial x^2} = 0 \tag{9.4}$$

where, t: time,

x: position,

c: celerity of the wave propagation.

Eq. (9.4) describes the propagation of a small disturbance on the surface of shallow body of ideal liquid in general. Assuming c = 1, one can obtain:

$$\frac{\partial^2 f}{\partial t^2} - \frac{\partial^2 f}{\partial x^2} = 0 \tag{9.5}$$

This equation has two real characteristics described by the equations dx/dt = 1 and dx/dt = −1, respectively.

9.3. Boundary and Initial Conditions

To solve a hyperbolic equation, it is necessary to specify both the boundary conditions and the initial conditions for the problem. Boundary conditions are conditions that must be satisfied at the boundaries of the region over which the equation is being solved. For example, in the case of a wave equation, the boundary conditions might specify the amplitude and phase of the wave at the boundaries of the region. In general, the number and form of the boundary conditions will depend on the specific problem being considered.

Initial conditions, on the other hand, specify the state of the system at the initial time. For example, in the case of a wave equation, the initial conditions might specify the displacement of the wave at a particular time. Like boundary conditions, the number and form of the initial conditions will depend on the specific problem being considered.

It is important to note that both the boundary conditions and the initial conditions are essential for uniquely determining the solution to a hyperbolic equation. Without these conditions, the equation will not have a unique solution.

We can consider Eq. (9.1), which is also known the advection equation, generally on a limited interval as $\partial f/\partial t + c\partial f/\partial x = 0$ for $t > 0$ and $0 < x < 1$.

We have to choose boundary conditions at x = 0 and x = 1. In hyperbolic problems, we can define inflow and outflow boundaries. However, the boundary condition must be given at inflow boundary, but the boundary condition can not be given at outflow boundary.

In order to solve the above equation, we need; for c > 0: one boundary condition at x = 0, and for c < 0: one boundary condition at x = 1.

If c > 0 the characteristic lines are directed to the right and if c < 0 they are directed to the left as can be seen from Figure 9.1. Looking the characteristic lines in Figure 9.1, we can see that the solution must be specified on the boundary at x is equal to 0, except from the initial data, in order to be defined for all time. Moreover, no data can be needed at the outflow boundary.

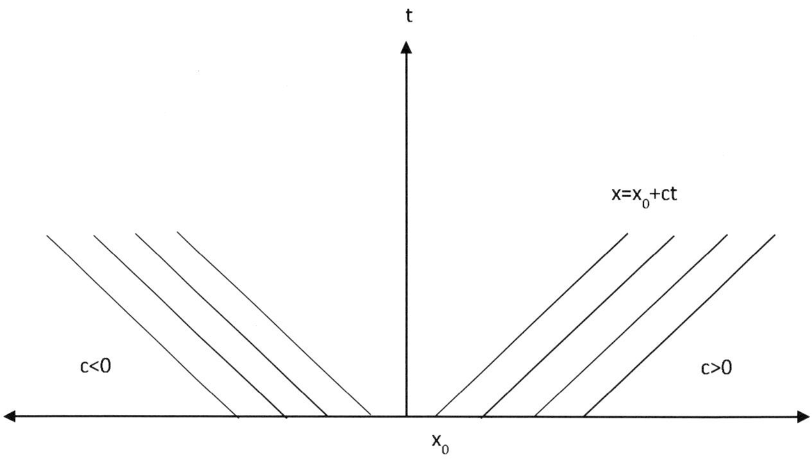

Figure 9.1. Characteristic lines for the advection equation.

We need also initial conditions. If we specify initial data for t = 0 as f(0, x) = f_0(x) and boundary data for x = 0 as u(t, 0) = g(t), then the solution can be obtained as u(t, x) = u_0(x − ct) if x − at > 0, g(t − c−1x) if x –ct < 0.

9.4. Review of Some Hyperbolic Equations

9.4.1. General

We want to consider some examples of the hyperbolic partial differential equations. One of them is the one-way wave equation as can be seen in Equation (9.5).

Another one is used in engineering as the convection equation

$$\frac{\partial f}{\partial t} + \frac{\partial f}{\partial x} = 0 \tag{9.6}$$

Both equations are same, mathematically.

2-Dimensional convection equation can be written as:

$$\frac{\partial f}{\partial t} + \frac{\partial f}{\partial x} + \frac{\partial f}{\partial y} = 0 \qquad (9.7)$$

9.4.2. The Advection Equation in One Dimension

We can consider the advection equation in one dimension, which is the first order of wave equation. The first order wave equation in one-dimensional space is given in Equation (9.1), where c is a constant, t represents time, and x represents the spatial variable. This is known as one-way wave equation with a constant value of c, which is the advection velocity.

At the initial stage, we take f(t, x), which we always assume to be 0. That is, for all real integers x, f(0, x) must be equal to a given function f_0, and we want to know the values of f(t, x) for positive values of f. This is referred to as an initial value problem. f(x, t) is subject to the initial and boundary conditions for $-\infty < x < \infty$.

$$f(x, 0) = f_0 \qquad (9.8)$$

The solution of the above equation for x gives a family of characteristics. These characteristics are straight lines shifted to the left in the x-t plane, if c is positive. The slopes of the characteristic lines are constant. These lines are inclined to the x-axis at an angle, θ, for $t \geq 0$

$$\theta = \tan^{-1}(1/c) \qquad (9.9)$$

The explicit solution is for positive c:

$$f(x, t) = f_0(x - ct) \qquad (9.10)$$

For negative c the solution is given as:

$$f(x ,t) = f_0(x + ct) \qquad (9.11)$$

9.4.3. The Advection Equation in One Dimension Using a Finite Difference Method

We examine the advection equation, Equation (9.1), using a finite difference method, which called one- dimension upwind scheme (Bermudez and Vazquez, 1994). The grid points of finite difference are shown in Figure 9.2.

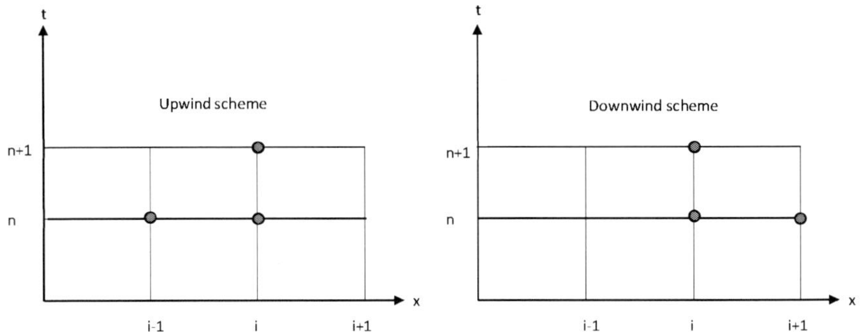

Figure 9.2. The stencil for upwind and downwind schemes.

Depending on which direction the solution of the equation (9.1) is to be translated, and therefore on the advection velocity c is positive or negative, two different finite difference algorithms can be obtained. These final algorithms are given in Equation (9.12) and Equation (9.13).

If $c > 0$, the FTBS which is an upwind scheme can be used:

$$f_i^{n+1} = f_i^n - c\Delta t/\Delta x\,(f_i^n - f_{i-1}^n) + O(\Delta t^2, \Delta x\Delta t) \quad (9.12)$$

If $c < 0$, FTFS which is a downwind scheme can be used:

$$f_i^{n+1} = f_i^n - c\Delta t/\Delta x\,(f_{i+1}^n - f_i^n) + O(\Delta t^2, \Delta x\Delta t) \quad (9.13)$$

in which $O(\Delta t^2, \Delta x\Delta t)$ is the order of accuracy.

9.4.4. The Wave Equation in Two Spatial Dimensions

Now, we'll apply the techniques we've learned so far to the case of a wave equation in two dimensions (Rezzolla, 2020).

$$\frac{\partial^2 f}{\partial t^2} = c^2 \frac{\partial^2 f}{\partial x^2} + \frac{\partial^2 f}{\partial y^2} \quad (9.14)$$

Similar to the one-dimensional example, the wave equation can also be solved by finding the solutions to a set of three first-order advection equations in this case.

$$\frac{\partial r}{\partial t} = c\frac{\partial s}{\partial x} \quad (9.15)$$

$$\frac{\partial I}{\partial t} = c\frac{\partial s}{\partial y} \quad (9.16)$$

$$\frac{\partial s}{\partial t} = c\frac{\partial r}{\partial x} + \frac{\partial l}{\partial y} \quad (9.17)$$

Once the following definitions have been made,

$$r = c\frac{\partial f}{\partial x} \quad (9.18)$$

$$I = c\frac{\partial f}{\partial y} \quad (9.19)$$

$$s = \frac{\partial f}{\partial t} \quad (9.20)$$

9.5. Finite Difference Schemes in Hyperbolic Differential Equation

9.5.1. Finite Difference Formulation for Schemes

There are several finite difference schemes used in the solution of hyperbolic partial differential equations (Szymkiewicz, 2010). In this part, the application of these schemes to the 1-D wave equation will be explained. This equation, Equation (9.1), also called the advection equation, is given below,

$$\frac{\partial f}{\partial t} + c\frac{\partial f}{\partial x} = 0$$

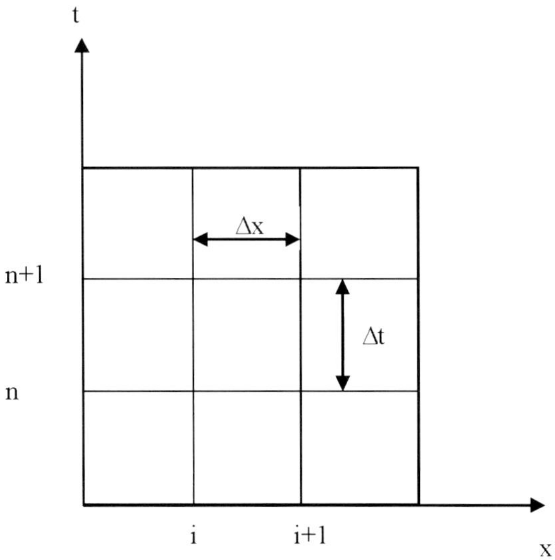

Figure 9.3. A Grid system for forward Euler scheme.

The schemes used in hyperbolic differential equations generally classified into two group as explicit and implicit schemes. The explicit schemes are: 1. Forward Euler scheme, 2. Upwind scheme, 3. Lax-Friedrichs scheme, 4. Lax-Wendroff scheme, 5. Leapfrog scheme, and 6. Fourth-order Leapfrog scheme. The implicit schemes are: 1. Backward Euler scheme, 2. Crank-Nicolson scheme, and 3. Box scheme.

The equivalent finite difference form of Equation (9.1) can be written for Forward Euler scheme as,

$$\frac{f_i^{n+1} - f_i^n}{\Delta t} + c\frac{f_{i+1}^n - f_{i-1}^n}{2\Delta x} = 0 \tag{9.21}$$

Rearranging this equation yields,

$$f_i^{n+1} = f_i^n + \frac{1}{2} r(f_{i+1}^n - f_{i-1}^n) \tag{9.22}$$

where,

$$r = c\frac{\Delta t}{\Delta x} \tag{9.23}$$

where Δx is the space interval, Δt is the time interval, c is the advection celerity.

Similarly, we can formulate the other schemes. Explicit and implicit finite difference schemes and their formulations are shown in Table 9.1 and Table 9.2, respectively.

Table 9.1. Explicit finite difference schemes for first order 1-D wave equation

Number of Scheme	Finite Difference Scheme	Scheme formulation
1	Forward Euler scheme	$f_i^{n+1} = f_i^n + \frac{1}{2}r(f_{i+1}^n - f_{i-1}^n)$
2	Upwind scheme	$f_i^{n+1} = f_i^n + r(f_{i+1}^n - f_i^n)$
3	Lax-Friedrichs scheme	$f_i^{n+1} = \frac{1}{2}(f_{i+1}^n - f_{i-1}^n) + \frac{1}{2}r(f_{i+1}^n - f_{i-1}^n)$
4	Lax-Wendroff scheme	$f_i^{n+1} = f_i^n + \frac{1}{2}r(f_{i+1}^n - f_{i-1}^n) + \frac{1}{2}r^2(f_{i+1}^n - 2f_i^n + f_{i-1}^n)$
5	Leapfrog scheme	$f_i^{n+1} = f_i^{n-1} + r(f_{i+1}^n - f_{i-1}^n)$
6	Fourth-order Leapfrog scheme	$f_i^{n+1} = f_i^{n-1} + \frac{4}{3}r(f_{i+1}^n - f_{i-1}^n) + \frac{1}{6}r(f_{i+2}^n - f_{i-2}^n)$

Table 9.2. Implicit finite difference schemes for first order 1-D wave equation

Number of scheme	Finite Difference Scheme	Scheme formulation
1	Backward Euler	$f_i^{n+1} = f_i^n + \frac{1}{2}r(f_{i+1}^{n+1} - f_{i-1}^{n+1})$
2	Crank-Nicolson	$f_i^{n+1} = f_i^n + \frac{1}{4}(f_{i+1}^n - f_{i-1}^n) + \frac{1}{2}r(f_{i+1}^{n+1} - f_{i-1}^{n+1})$
3	Box	$(1 - r)f_i^{n+1} + (1 - r)f_{i+1}^{n+1} = (1 - r)f_i^n + (1 + r)f_{i+1}^n$

9.5.2. The Courant-Friedrichs-Levy Stability Condition

Courant, Friedrichs and Levy, these 3 researchers published an article in 1928, working on partial differential equations. Although this article is written before the innovation of computers, it determines the convergence condition of any scheme in finite difference calculations. This condition is known as the Courant- Friedrichs-Levy condition or CFL requirement.

In the 1-D advection equation, which is the simple one of hyperbolic partial differential equations, the CFL requirement for the forward Euler scheme is as follows,

$$r = |c|\frac{\Delta t}{\Delta x} \leq 1 \tag{9.24}$$

This inequality is also known CFL stability condition. There are two velocity definitions in this expression which are wave propagation velocity, c, and numerical integration velocity $\Delta x/\Delta t$. Here, the idea is physical velocity should not exceed the integration velocity, namely, $|c|$ should be less than or equal to $\Delta x/\Delta t$. So, time interval Δt should be selected according to a given Δx value.

Most of the schemes used in hyperbolic partial differential equations are subject to stability condition. The CFL stability conditions and the order of accuracies are shown in Table 9.3 for all schemes used in 1-D advection equation.

Table 9.3. Stability and order of accuracy for finite difference schemes

Number of scheme	Finite Difference Scheme	Order of accuracy	CFL stability condition	Exact stability condition
1	Forward Euler	1	$r \leq 1$	Unstable
2	Upwind Scheme	1	$r \leq 1$	$r \leq 1$
3	Lax-Friedrichs Scheme	1	$r \leq 1$	$r \leq 1$
4	Lax-Wendroff Scheme	2	$r \leq 1$	$r \leq 1$
5	Leapfrog Scheme	2	$r \leq 1$	$r \leq 1$
6	4th-order Leapfrog Scheme	2	$r \leq 2$	$r \leq 0.728$
7	Backward Euler Scheme	1	None	None
8	Crank-Nicolson Scheme	2	None	None
9	Box Scheme	2	None	None

9.6. Applications of Hyperbolic Equations

9.6.1. Solution of a Linear Kinematic Wave Model

a. Using the following data derive a linear kinematic wave equation.
b. Determine Q_{i+1}^{n+1} using finite difference scheme for a linear kinematic wave model. Assume the lateral flow to be zero.

Given Data: A rectangular channel, 20 m wide, with bed slope of 0.01 and Manning's n = 0.040. The following flow rates are given: Q_i^{n+1} = 21 m^3/s, Q_i^n = 16 m^3/s and Q_{i+1}^n = 15 m^3/s. Take Δx = 1000 m and Δt = 6 min.

Solution: a. The equations for linear kinematic wave model with no lateral flow are given below.

Continuity equation,

$$\frac{\partial Q}{\partial x} + \frac{\partial A}{\partial t} = 0 \tag{9.25}$$

The kinematic wave approximation based on the following relationship,

$$A = aQ^b \tag{9.26}$$

This is a simplified momentum equation (see Chapter 10).

Using the above equation, we can obtain the kinematic wave equation as:

$$\frac{\partial Q}{\partial x} + abQ^{b-1}\frac{\partial Q}{\partial t} = 0 \tag{9.27}$$

This is a hyperbolic equation. The equivalent finite difference equation can be found using following terms:

$$Q = \frac{1}{2}(Q_{i+1}^{n} + Q_{i}^{n+1}) \tag{9.28}$$

$$\frac{\partial Q}{\partial x} = (Q_{i+1}^{n+1} + Q_{i}^{n+1})\,/\Delta x \tag{9.29}$$

$$\frac{\partial Q}{\partial t} = (Q_{i+1}^{n+1} + Q_{i+1}^{n})\,/\Delta t \tag{9.30}$$

Determine Q_{i+1}^{n+1} using finite difference scheme for a linear kinematic wave model: Assume lateral flow to be zero.

Solution: The finite difference form of this equation can be written as

$$Q_{i+1}^{n+1} = \frac{(\Delta t/\Delta x)Q_{i}^{n+1} + abQ_{i+1}^{n}\left(\frac{Q_{i+1}^{n} + Q_{i}^{n+1}}{2}\right)^{b-1}}{\Delta t/\Delta x + ab\left(\frac{Q_{i+1}^{n} + Q_{i}^{n+1}}{2}\right)^{b-1}} \tag{9.31}$$

b. Determination of Q_{i+1}^{n+1}:

a and b can be obtained from Manning equation as

$$Q = \frac{1}{n} R^{2/3} S^{1/2} A \tag{9.32}$$

where, Q is the discharge,

n is the Manning's friction coefficient,

R is the hydraulic radius, as R = A/P,

P is the wetted perimeter,

S is the channel slope,

A is the wetted area.

We can solve A as,

$$A = [(nP^{2/3}/(S^{1/2})]^{0,6} Q^{0,6} \tag{9.33}$$

This equation shows the equivalent of Equation (9.27) as $A = aQ^b$, which is obtained using Manning's friction formula.

The wetted perimeter is approximately equal to width of channel. For P = B, using the given data we can calculate,

$$a = (nB^{2/3}/(S^{1/2}))^{0.6} = 1.91$$

$$b = 0.6.$$

Using the given data, from Equation (9.31) we can calculate Q_{i+1}^{n+1}

$Q_{i+1}^{n+1} = [(360/1000)\times 21 + 1.91\times 0.6x15x((21 + 15)/2)^{-0.4}]/[360/1000 + 1.91\times 0.6((21 + 15)/2)^{-0.4}] = 18\ m^3/s.$

9.6.2. Example for Flood Routing Using Kinematic Wave Model

For the following data determine the flow routing using the linear kinematic wave model.

Given data: A rectangular channel, 80 m wide and 4 km long with bed slope of 0.0169 and Manning's n = 0.03.

The inflow hydrograph for the channel is given in Table 9.4.

Table 9.4. Inflow hydrograph values

Time (min)	0	4	8	12	16	20	24
Inflow (m^3/s)	6	10	18	24	30	28	23

The initial condition is a uniform flow as constant discharge as 6 m^3/s.

Use the linear kinematic wave model to route the inflow hydrograph through the channel.

Take Δx = 1000 m and Δt = 4 min. There is no lateral inflow.

The equation for routing is derived in Equation (9.31) as

Q_{i+1}^{n+1}= [($\Delta t/\Delta x$)Q_i^{n+1} + ab(Q_{i+1}^{n})((Q_{i+1}^{n} + Q_i^{n+1})/2)$^{b-1}$]/[$\Delta t/\Delta x$ + ab((Q_{i+1}^{n} + Q_i^{n+1})/2)$^{b-1}$]

Wetted perimeter can be taken as approximately equal to width of the channel.

Solution: Width of channel (B) = 80 m; Bed slope (S) = 0.0169; Manning's n = 0.03.

Considering Equation (9.32), a and b can be obtained from Equation (9.3) and Equation (9.35) using given data,

$a = (nB^{2/3}/(S^{1/2}))^{0.6} = (0.03\times80^{2/3}/(0.0169)^{0.5})^{0.6} = 2.39$

b = 0.6.

Computation of outflow hydrograph: For i = 1 all values are given for all t. For n = 1, values are given for all i as 6 m^3/s.

i = 1, n = 1: Q_i^{n+1}=10 m^3/s, Q_i^n = 6 m^3/s and Q_{i+1}^n = 6 m^3 /s.

For Δx = 1000 m, and Δt = 4 min = 240 s, we can calculate outflow discharge as,

Q_{i+1}^{n+1} = [($\Delta t/\Delta x$) Q_i^{n+1} +ab(Q_{i+1}^{n})(Q_{i+1}^{n} + $Q_i^{n+1}$$Q_i$)/2)$^{b-1}$]/[$\Delta t/\Delta x$ +ab((Q_{i+1}^{n} + Q_i^{n+1})/2)$^{b-1}$]

= [(240/1000)×10 + 2.39×0.60x6((6 +10)/2)$^{-0.4}$]/[(240/1000) + 2.39×0.60((6 +10/2)$^{-0.4}$)] = 7.11 m^3/s.

Using Excel, the other values of outflow are computed in a similar way and shown in Table 9.5. Initial inflow hydrograph values taking from Table 9.4, and computed values of outflow taking from Table 9.5 at the end of channel (for distance 4000 m) are shown in Figure 9.4.

Table 9.5. Calculated discharge values

Time (min)	Time index (n)	Distance (m)					Remark
		0	1000	2000	3000	4000	
		i = 1	i = 2	i = 3	i = 4	i = 5	
0	1	6	6	6	6	6	Given
4	2	10	7.11	6.29	6.07	6.02	Computed
8	3	18	10.54	7.49	6.45	6.13	Computed
12	4	24	15.17	9.84	7.40	6.47	Computed
16	5	30	20.62	13.42	9.20	7.22	Computed
20	6	28	23.39	16.90	11.66	8.51	Computed
24	7	23	23.25	19.17	14.16	10.24	Computed

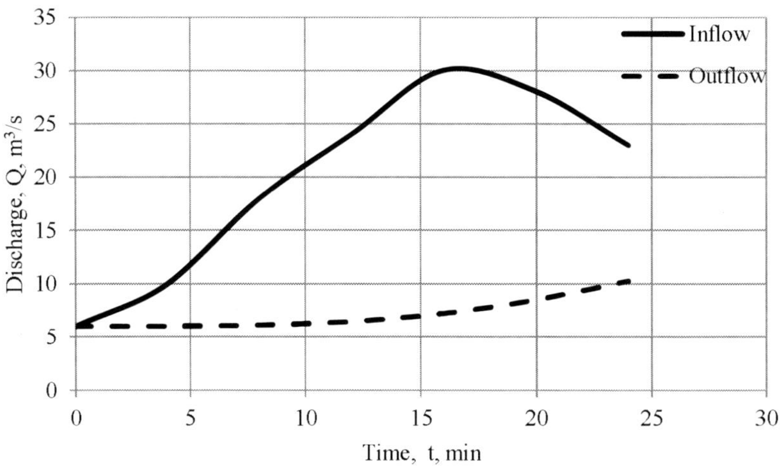

Figure 9.4. Inflow and outflow hydrographs.

9.6.3. Determination of Stability Criterion

For the following data, determine the stability criterion of kinematic wave model.

Given data: A rectangular channel with a bed slope of 0.0144 and Manning's n = 0.025 is 70 m wide. A segment of the channel has a discharge of 120 m^3/s. For kinematic wave routing, the space interval $\Delta x = 1000$ m is used. Assume that the hydraulic radius equals the flow depth.

Solution: Applying the Courant stability condition to the kinematic wave solution will allow us to determine the maximum value of the time step.

Computation of flow depth: Consider Manning's equation, using n = 0.025; S = 0.0144; B = 70 m; R ≈ y,

$$Q = \frac{1}{n} R^{2/3} S^{1/2} A$$

$$120 = (1/0.025)y^{2/3}(0.0144)^{0.5} \times 70y$$

Solving for y,

$$y = 0.54 \text{ m}$$

Calculation of kinematic wave celerity: Velocity in a natural channel, for a given area A can be written as, velocity: V = Q/A. In a natural channel for a given depth y can be written the kinematic wave celerity as

$$C_k = (dQ/dA) = (1/B)(dQ/dy)$$

where V is the velocity (m/s), Q is the flow (m^3/s), C_k is the kinematic celerity (m/s), A is the flow area (m^2), B is the top width of flow (m). Using the given data, we can calculate the kinematic wave celerity,

$$C_k = (1/B) \times d((1/n)\, R^{2/3}\, S^{1/2}\, A)/dy$$

$$C_k = (1/70) \times d((1/0.025) \times y^{2/3} \times (0.0144)^{1/2} \times y \times 70)/dy$$

$$C_k = 4.8 \times d(y^{5/3})/dy = 4.8 \times 3/5 x y^{2/3}$$

$$C_k = 4.8 \times 3/5 \times 0.54^{2/3} = 1.91 \text{ m/s}$$

Stability condition: Courant condition is known as: $C_r < 1$

$$C_k/(\Delta x/\Delta t) < 1$$

$$\Delta t < \Delta x/(C_k) = 1000/1.91$$

To satisfy the stability condition, the time interval should be,
$\Delta t < 524$ s = 8.7 min.

9.6.4. Example for Numerical Solution of an Overland Flow

Using kinematic wave theory, calculate the depth of water on a watershed for the following data at points (j,n) as h_2^2, h_2^3, h_2^4, and h_3^2. Consider Figure 9.5 and Figure 9.6.

Given data: Watershed length: L = 144 m, Watershed slope: S_o= 0.002, Average width: B = 0.9 m, excess rainfall i =19 cm/hour and Rainfall duration: T = 8 minutes. Kinematic parameters are found as, α = 15.72 and β = 1.71.

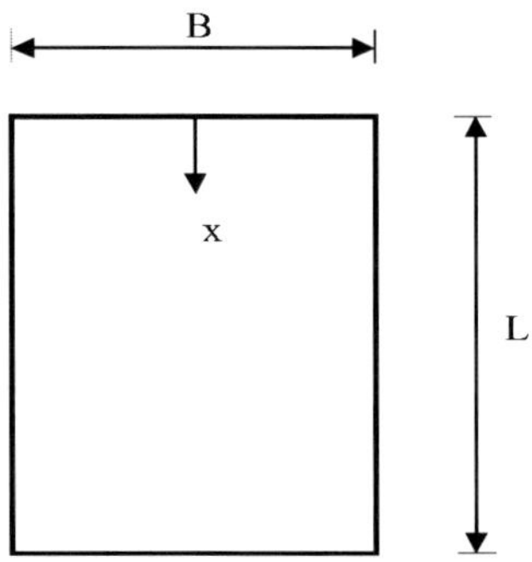

Figure 9.5. Watershed plan.

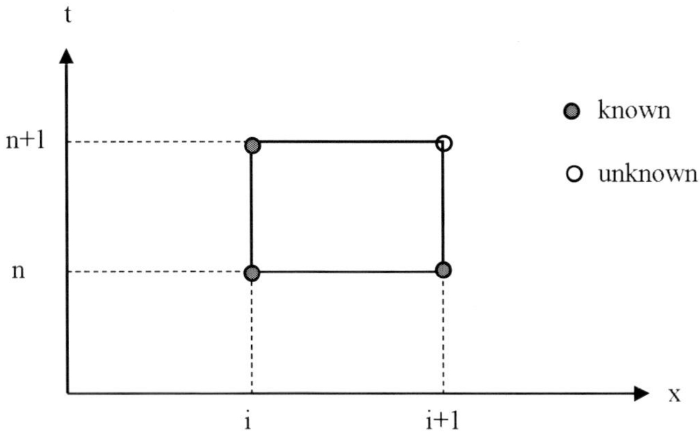

Figure 9.6. Grid system for 4 points.

Solution: Kinematic wave equation for rectangular watershed geometry can be derived as

$$\frac{\partial h}{\partial t} + \alpha\beta h^{\beta-1}\frac{\partial h}{\partial x} = I$$

where h is the water depth, and i is rainfall excess. This equation is derived in Chapter 4, in the title of Kinematic Wave Model. In this equation, q = Q/B = αh^{β}, where q is the rate flow per unit width. We obtain the following equation using forward in time and centered in space finite differences

$$\frac{h_j^{n+1} - h_j^n}{\Delta t} + \alpha\,\beta\left(h_j^n\right)^{\beta-1}\left(\frac{h_{j+1}^n - h_{j-1}^n}{2\Delta x}\right) = I$$

where Δx and Δt are space and time intervals, respectively.

We obtain the following equation,

$$h_j^{n+1} = h_j^n + \Delta t\left[i - \alpha\beta(h_j^n)^{\beta-1}\frac{h_{j+1}^n - h_{j-1}^n}{2\Delta x}\right]$$

Initial and boundary conditions: There is no rainfall at the upper part of the watershed,

$h_j^1 = 0$ (initial condition)

$h_1^n = 0$ (boundary condition)

Time and space intervals are selected, respectively as $\Delta t = 1$ s, and $\Delta x = 16$ m.

Using the given data we can calculate the flow depth for time steps and locations as:

$$h_2^2 = h_1^1 + 1\left[\frac{0.19}{3600} - (15.72 \times 1.71)(h_2^1)^{0.71}\frac{h_3^1 - h_1^1}{2\times 16}\right]$$

$$h_2^2 = 0 + 1\frac{0.19}{3600} = 0.000053 \text{ m}$$

$$h_3^2 = h_4^2 = h_5^2 = \cdots = 0.000053 \text{ m}$$

$$h_2^3 = 0.000053 + \left[\frac{0.19}{3600} - (15.72) \times \left(\frac{0.000053}{1}\right)^{1.71-1}\frac{(h_3^2 - h_1^2)}{2\times 16}\right] =$$

0.00106 m

9.6.5. Numerical Solution of Shoreline Change

Calculate the shoreline change induced by a groin using a one-line model. A length of 900 m of shoreline will be considered for the coastal alteration. The study area is located in the Efirli coastline region of the Central Black Sea Region, west of the Ordu province, at 41°010567"N and 37°810104"E. (Figure 9.7). Table 9.6 describes the wave climate in the study area. Consider NE is the dominating wave direction.

Table 9.6. Wave characteristics of study area

Wave direction	Wave angle with shoreline	Average wave height	Significant wave height	Wave period	Wave length	Wave height at breaking	Water depth at breaking	Breaking angle	Angle with x axis
	Degree	m	m	s	m	m	m	Degree	Degree
		H_0	H_s	T_b	L_0	H_b	h_b		
NE	44.0	0.33	0.60	2.8	12	0.45	0.58	22.29	9.71

Figure 9.7. Coastal area of Ordu Province near Efirli.

A one-line model is used to develop the numerical solution (Hanson, and Kraus, 1989). The shape of any beach cross-shore profile is assumed to be in balance, and all seabed contours are generally parallel. The depth of closure, below which sediment motion is minimal, is a depth limit that limits cross-shore flow. The longshore sediment transport equation and the sand continuity equation were combined to generate this model (Kamphuis, 2000). The sand continuity equation that describes the coastline change is given in the following equation.

$$\frac{\partial Q}{\partial x} + \frac{\partial A}{\partial t} + q = 0 \quad (9.34)$$

where Q is the sediment transport rate along the shore line (m^3/s), A is the sediment transport area (m^2), q is the source and/or sink along the coast ($m^3/s/m$), x and t are the space and time coordinates. If there is no source and or sink along the coast, q will be zero. For q = 0, it reduces the following equation for the coordinate system given in Figure 9.8;

$$\frac{\partial y}{\partial t} = -\left[\frac{1}{D_B + D_C}\right]\frac{\partial Q}{\partial x} \tag{9.35}$$

where D_B is the beach berm height above still water level,

D_C is the depth of closure.

To solve Equation 9.35, the initial shoreline position over the full length, boundary conditions on each end of the beach, and values for Q, D_B, and D_C must be given.

The selected empirical equations determine the amplitude of the longshore transport rate, Q_0. The breaking wave angle is defined as the point at which a wave breaks, where α_b is the angle formed by the breaking wave crests with respect to an axis that runs parallel to the shoreline's trend and dy/dx is the orientation of the shoreline in the area or shoreline gradient.

The longshore sediment transport rate, Q is expressed by the following equations as in Equation (9.36) (SPM, 1984).

$$Q = Q_0 \sin 2\alpha_{bs} \tag{9.36}$$

where α_{bs} = angle of breaking waves to the local shoreline as can be seen in Figure 9.9. The angle between the breaking wave crests and the shoreline can be expressed as the difference between, α_s the shoreline orientation, and α_b the breaking wave angle.

$$\alpha_{bs} = \alpha_b - \alpha_s \tag{9.37}$$

where α_b is the angle of breaking with respect to x-axis, α_s is the angle that shoreline makes with x-axis and α_{bs} is the angle of breaking with respect to shoreline as seen in Figure 9.9.

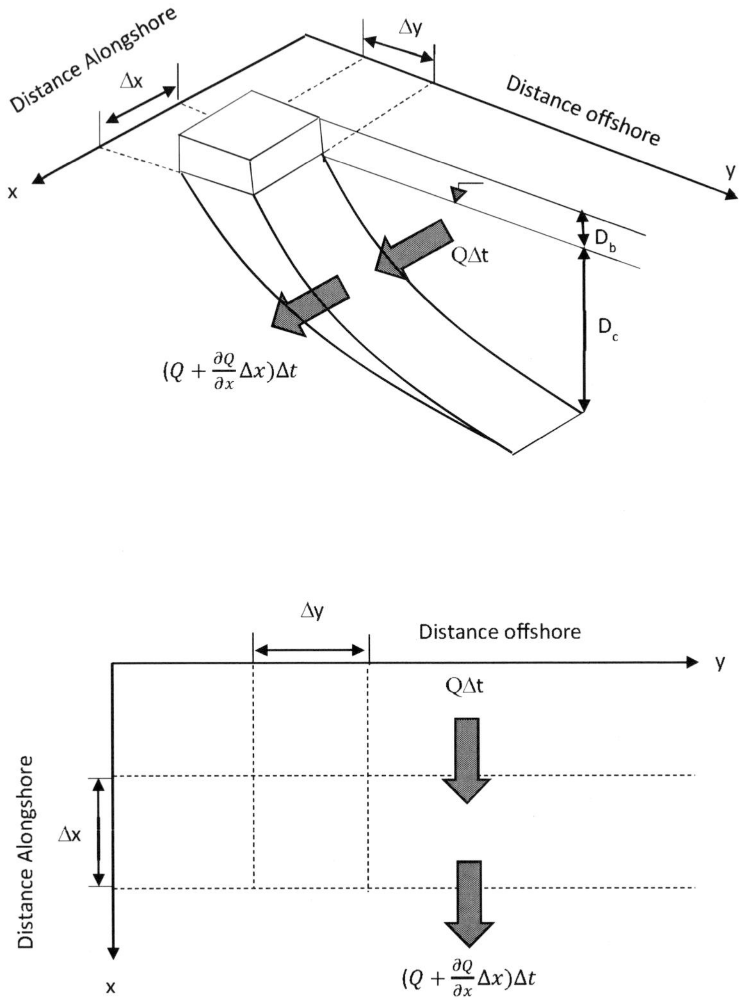

Figure 9.8. Definition sketch for shoreline change calculation.

The shoreline orientation, α_s can be expressed as the change in the local shoreline position, as shown in Figure 9.9.

$$\alpha_s = \tan^{-1}(dy/dx) \tag{9.38}$$

$$\alpha_{bs} = \alpha_b - \tan^{-1}(dy/dx) \tag{9.39}$$

where α_b angle to normal x direction

$$Q_0 = K'(H^2C_g)_b = K'H_b^2C_{gb} \qquad (9.40)$$

in which H = wave height, C_g = wave group speed given by linear wave theory, b = subscript denoting wave breaking condition.

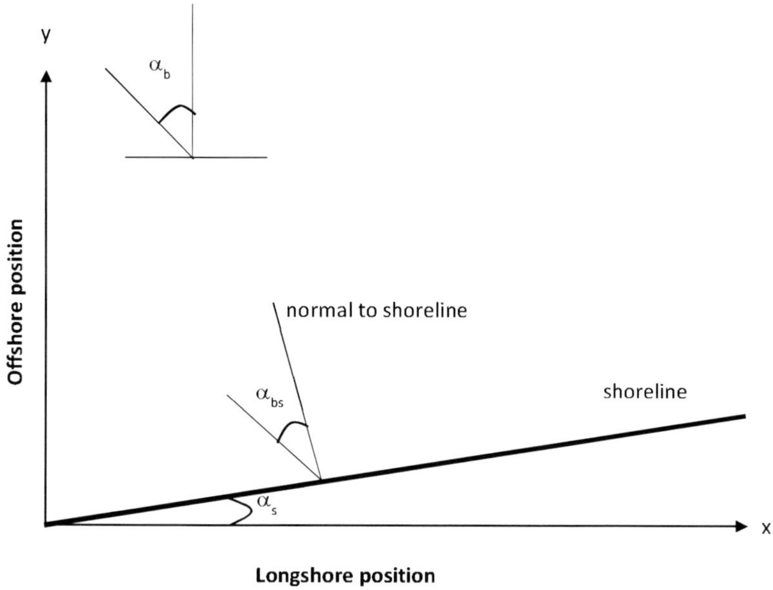

Figure 9.9. Definition of breaking wave angles.

In Equation (9.40), K' can be written as

$$K' = K/[16(s-1)(p-1)(r)^{5/2}] \qquad (9.41)$$

in which K is the dimensionless empirical coefficient (about 0.4),

s is the ratio of density of sand (2.65 t/m^3) to density of sea water (1.03 t/m^3),

p is the porosity of sediment (about 0.4),

r is the conversion coefficient (as 1.416).

Wave group speed can be obtained by the following equation:

$$(C_g)_b = [g(H_b/\gamma_b)]^{0.5} = (gh_b)^{0.5} \qquad (9.42)$$

g is the gravitational acceleration, m/s^2,

H_b is the wave height at breaking, m,

h_b is the depth of water at breaking, m,

γ_b is the breaker index (taken as 0.78).

Equation (9.35) and Equation (9.36) can be solved by any numerical or analytical method subject to initial and boundary conditions.

Considering Figure 9.8 the forward in time and forward in space scheme is applied as follows,

$$y_i^{n+1} = y_i^n - (\Delta t/\Delta x)(1/D)(Q_{i+1}^n - Q_i^n) \tag{9.43}$$

and

$$Q = Q_0(\text{Sin } 2(\alpha_b - \tan^{-1}(dy/dx))) \tag{9.44}$$

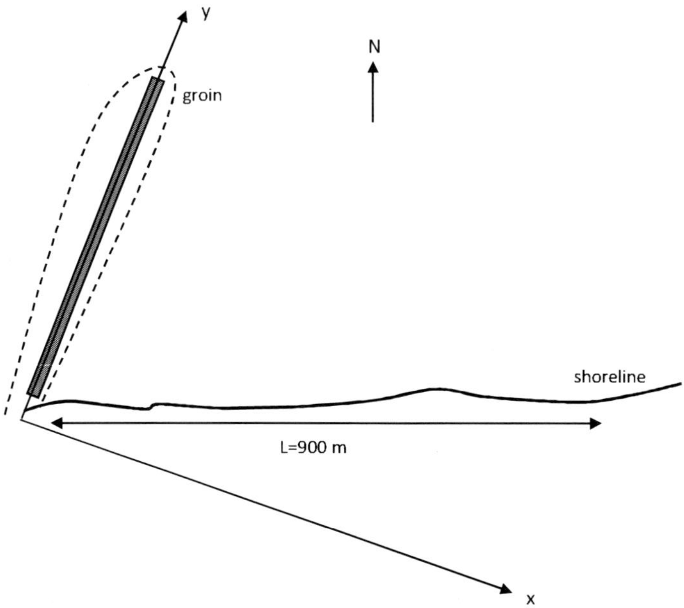

Figure 9.10. Shoreline and selected coordinate system.

Model stability in the explicit scheme solution should be carefully considered. This stability condition is used to establish the necessary grid spacing and time step. The stability criterion for an explicit scheme can be expressed as:

$$½ > R_s = 2xK'\Delta t(H^2\,C_g)_b/(D(\Delta x^2) \tag{9.45}$$

In order to start calculation, the y axis of coordinate system is selected as it passes through the center of groin (Figure 9.10).

Using H_b = 0.45 m, D =2.362 m, Δt = 24 hours, Δx = 100 m, it is possible to calculate following parameters.

$K' = K/[16(s-1)(p-1)(1/r)^{5/2} = 0.4/[16x(2.65 - 1.03)\ (1 - 0.4)(1.416)^{2.5}] = 0.011$

and

$C_{gb} = (gH_b/\gamma_b) = (9.81x0.45/0.78)^{0.5} = 2.38$

Stability condition of the explicit scheme can be calculated as:

$R_s = 2*K'*\Delta t*H_b{}^2*C_{gb}/(D*\Delta x^2) =$

$= 2x0.011x24x3600x0.45x0.45x2.38/(2.362x100x100) = 0.04 < 0.5$

The other parameters:

$\Delta t/(D*\Delta x) = 24x3600/(2.362x100) = 365.7917$

$Q_0 = K'(H^2\ C_g)_b = 0.011x0.45x0.45x2.38 = 0.0052$

$Q = Q_0(Sin2(\alpha_b - \tan^{-1}(dy/dx)))$

To calculate the shore change, we need to solve the following two equations, simultaneously,

$Q = 0.0052x\sin(22.29 - \tan^{-1}(\Delta y/\Delta x))$

$y_i{}^{n+1} = y_i{}^n + 365.7917(Q_i{}^n - Q_{i+1}{}^n)$

The boundary conditions for the two ends of the model should be specified. It is convenient to use a constant sediment transport condition at the ends of the model.

$Q_i = Q = \text{Constant}$

This condition states that the sediment transport rate at the model's endpoints remains constant regardless of the conditions inside the model. The effect of a structure that is long enough to block the passage of sediment is described by the other boundary condition for the model's beginning side.

$$Q_i = 0$$

For initial conditions it is necessary to know the shoreline position at the beginning. The initial conditions are shown in Table 9.7.

The result of the computations are given for shoreline change in Table 9.8.

The computed longshore profiles are included in Table 9.9.

Table 9.7. Initial conditions

Distance (m)	0	100	200	300	400	500	600	700	800	900
Wave height (m)	0.60	0.60	0.60	0.60	0.60	0.60	0.60	0.60	0.60	0.60
Wave angle (^{0})	44.0	44.0	44.0	44.0	44.0	44.0	44.0	44.0	44.0	44.0
Shoreline position (m)	0.0	62.5	125.0	187.5	250.0	512.5	375.0	437.5	500.0	562.5
Longshore transport	0.00	0.00	0.00	0.00	0.00	0.00	0.00	0.00	0.00	0.00

Table 9.8. Computed Shoreline change

Distance (m)	0.00	62.50	125.00	187.50	250.00	312.50	375.00	437.50	500.00	562.50
1 Year	35.77	82.78	135.31	192.17	251.89	313.19	375.22	437.56	500.00	562.50
2 Year	54.86	100.10	149.44	202.50	258.66	317.18	377.32	438.43	500.00	562.50
3 Year	69.53	113.97	161.81	212.84	266.69	322.91	380.93	440.17	500.00	562.50
4 years	81.85	125.80	172.68	222.33	274.52	328.86	384.92	442.16	500.00	562.50
5 years	92.56	136.14	182.29	230.88	281.70	334.45	388.74	444.09	500.00	562.50
6 years	102.87	146.13	191.61	239.22	288.76	339.98	392.54	446.03	500.00	562.50

Table 9.9. Computed Longshore profiles

Distance (m)	0	100	200	300	400	500	600	700	800	900
Shoreline	0.00	0.00	0.00	0.00	0.00	0.00	0.00	0.00	0.00	0.00
1 Year	35.77	20.28	10.31	4.67	1.89	0.69	0.22	0.06	0.00	0.00
2 Years	54.86	37.60	24.44	15.00	8.66	4.68	2.32	0.93	0.00	0.00
3 years	69.53	51.47	36.81	25.34	16.69	10.41	5.93	2.67	0.00	0.00
4 years	81.85	63.30	47.68	34.83	24.52	16.36	9.92	4.66	0.00	0.00
5 years	92.56	73.64	57.29	43.38	31.70	21.95	13.74	6.59	0.00	0.00
6 yeas	102.87	83.63	66.61	51.72	38.76	27.48	17.54	8.53	0.00	0.00

The computed shoreline profiles for different times are shown in Figure 9.11.

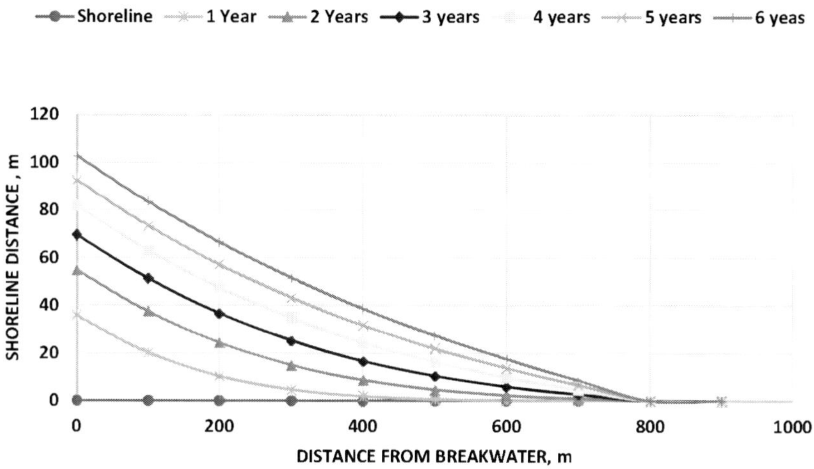

Figure 9.11. Computed shoreline profiles for different times.

Chapter 10

Applications of Unsteady Open Channel Flows

10.1. Introduction

This chapter discusses the applicability of approximate solutions to unstable open channel flow. The complete equations of open channel flow (the Saint Venant equations) can be simplified for some situations without fundamentally changing the character of the solution. As a result, the question of what criteria must be met for the reduced models to be used successfully instead of the whole model equations emerges.

Some examples of open channel flow are as:

- Rivers, streams, creeks, and natural channels.
- Storm runoff, flood waves, tides, and tsunamis.
- Water supply channels, sewer lines, and street gutters.
- Canals, culverts, aqueducts, and drainage ditches
- Weirs and gates.
- Water tunnels, irrigation canals.
- Flow in reservoirs and lakes.
- Flow in estuaries.
- Overland flow.
- Flow on surface irrigation.
- Flow on parking lots.
- Flow on roofs.

10.2. The Saint-Venant Equations

The 1-D Saint-Venant equations depend on the principles of mass and momentum conservation in unsteady open channel flow (Chow, 1959; Chaudhry, 2008). The continuity equation for section without lateral inflow can be written as

$$V\frac{\partial y}{\partial x}+y\frac{\partial V}{\partial x}+\frac{\partial y}{\partial t}=0 \tag{10.1}$$

The momentum equation can be written as

$$\frac{\partial V}{\partial t} + V\frac{\partial V}{\partial x} + g\frac{\partial y}{\partial x} - g(S_0 - S_f) = 0 \qquad (10.2)$$

In Equation (10.1) and Equation (10.2), V is the mean velocity, y is the flow depth, g is the acceleration of gravity, S_f is the friction slope, S_o is the channel bottom slope, x is the distance, and t is the time. Figure 10.1 shows the definition of slopes as S_o and S_f, in open channel flow. Lateral inflow has been neglected in Equation (10.1) and Equation (10.2).

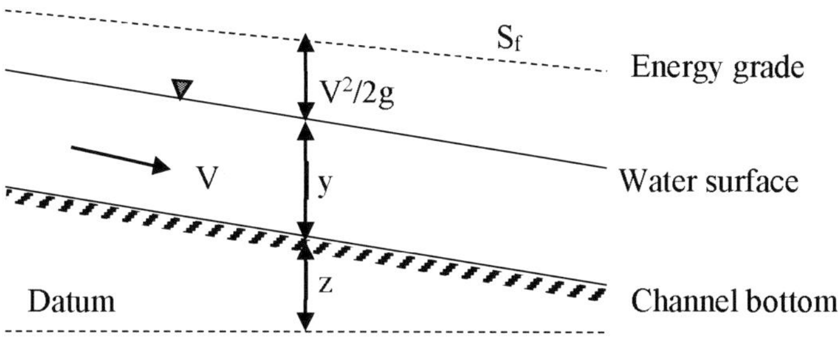

Figure 10.1. Definition of slopes in open channel flow.

In Equation (10.1), the third term is called the rate-of-rise, the second is the prism storage, and the first is the wedge storage term. In Equation (10.2), the first term is denoted the local acceleration, the second is the convective acceleration, the third is the pressure gradient, the fourth is the bottom slope, and the fifth is the friction slope terms (Cunge et al., 1980; Chaudhry, 2008).

10.3. Open Channel Flow Models

10.3.1. Steady Flow

In open channel flow there are many different types of model depending on approximations. In steady flow, the rate-of-rise term in Equation (10.1) and local acceleration term in Equation (10.2) are zero. The combination of Equation (10.1) and Equation (10.2) without those two terms leads to the steady flow (or backwater) equation

$$\frac{dy}{dx} = \frac{S_0 - S_f}{1 - F^2} \tag{10.3}$$

in which F is the Froude number and is defined,

$$F = \frac{V}{(gy)^{0.5}} \tag{10.4}$$

10.3.2. Quasi-Steady Approximation

This approximation to the unsteady flow equations consists of neglecting the local acceleration term in Equation (10.2). This is done on the ground that this is usually very small compared to the remaining terms in Equation (10.2). The unsteady element of the problem is preserved through the rate-of-rise term.

10.3.3. Kinematic and Diffusion Wave Models

The kinematic and diffusion wave models are approximations to the dynamic wave model expressed by the complete Saint-Venant equations. The diffusion wave model assumes that the inertia terms in Equation (10.2) (local and convective acceleration terms) are negligible as compared with the pressure gradient, friction and bottom slope terms. The kinematic model assumes that the inertia and pressure gradient terms are negligible as compared with the friction and bottom slope terms (Miller, 1984).

Strictly speaking, the kinematic and diffusion models are kinematic, in essence, since they are both based on the neglect of inertia term. The difference between them is in the wave properties; the kinematic wave is non-diffusive (it provides wave translation but not attenuation), while the diffusion model, due to the presence of the pressure gradient term, provides both wave translation and attenuation.

10.3.4. Inertial and Dynamic Wave Models

The inertial wave is a wave governed by inertia and the pressure gradient, to the exclusion of friction and bottom slope. This wave arises in connection with short wavelength phenomena in open channel flow.

The various models described here are summarized in Table 10.1.

Table 10.1. Summary of open channel flow models

Terms Neglected	Steady	Unsteady Quasi-Steady	Kinematic	Diffusion	Dynamic	Inertial
Rate-of-Rise	Yes*	No	No	No	No	No
Prism Storage	No	No	No	No	No	No
Wedge Storage	No	No	No	No	No	No
Local Acceleration	Yes	Yes	Yes	Yes	No	No
Convective Acceleration	No	No	Yes	Yes	No	No
Pressure Gradient	No	No	Yes	No	No	No
Friction Slope	No	No	No	No	No	Yes
Bottom Slope	No	No	No	No	No	Yes

* Yes means this term is neglected.

10.4. Applicability of Kinematic and Diffusion Wave Models

The criteria for the applicability of kinematic and diffusion wave models has been given by Ponce and Simons (1977). According to them, the error due to the use of the kinematic wave model is within 5% (measured in one period of propagation) provided the following inequality is satisfied in metric units,

$$\frac{TS_0u_0}{d_0} > 171 \tag{10.5}$$

in which T = wave period (akin to wave duration, second), S_o = bottom slope, u_o = mean velocity (m/s), and d_o = mean flow depth (m).

The error due to the use of the diffusion wave model is within 5%, (measured in one period of propagation) provided the following inequality is satisfied

$$TS_0\left(\frac{g}{d_o}\right)^{0.5} > 30 \tag{10.6}$$

in which g is the acceleration of gravity (9.81 m/s^2).

For instance, assume S_0 = 0.02, u = 1.1 m/s and d_o = 0.4 m. The wave period for the kinematic model to apply is T = 51.8 min; the wave period for the diffusion model is T = 5.0 min. The diffusion model is demonstrated to be more applicable than the kinematic model for a wider variety of slopes and wave periods. For flow on very steep slopes, however, the kinematic wave model will be sufficient. This is the reason for the use of the kinematic wave model in solution of overland flow problems (Ponce and Simons, 1977).

10.5. Unsteady Flow Models in Open Channel Flow

10.5.1. Summary of Saint-Venant Equations

Continuity Equations, in conservative form (Ligget, 1975),

$$\frac{\partial Q}{\partial x} + \frac{\partial A}{\partial t} = 0 \tag{10.7}$$

in which Q is the discharge, and A is the cross-section area of the flow. On the other hand, the non-conservative form of the continuity equation was written in Equation (10.1)

Momentum Equations in conservative form is,

$$\frac{1}{A}\frac{\partial Q}{\partial t} + \frac{1}{A}\frac{\partial\left(\frac{Q2}{A}\right)}{\partial x} + g\frac{\partial y}{\partial x} - g(s_0 - s_f) = 0 \tag{10.8}$$

Momentum equation in non-conservative form was written in Equation (10.2).

10.5.2. Kinematic Wave Model

Continuity Equation with lateral inflow, q can be written as:

$$\frac{\partial Q}{\partial x} + \frac{\partial A}{\partial t} = q \tag{10.9}$$

Simplified Momentum Equation: Using Manning's formula one can write

$$Q = \frac{1}{n} R^{\frac{2}{3}} S_0^{\frac{1}{2}} A \tag{10.10}$$

where Q is the flow rate, n is the Manning's friction coefficient, R is the hydraulic radius, S_o is the bottom slope of the channel, A is the wetted area.

In kinematic wave model, assuming $S_o \approx S_f$, we can write

$$A = \alpha Q^{\beta} \tag{10.11}$$

where α and β are the kinematic wave parameters.

On the other hand, from Equation (10.10), one can write

$$Q = \frac{1}{n}\left(\frac{A}{P}\right)^{\frac{2}{3}} S_0^{\frac{1}{2}} A \tag{10.12}$$

Solving A we can obtain:

$$A = [n^{0.6}\, P^{0.4}/S_0^{0.3}]\; Q^{0.6} \tag{10.13}$$

For wide channels, the wetted perimeter, P, is assumed to be equal to the channel width, B.

In Equation 10.11, the kinematic parameters are defined as

$$\alpha = n^{0.6}\, P^{0.4}/S_0^{0.3} \tag{10.14}$$

$$\beta = 0.6 \tag{10.15}$$

Derivation of Equation (10.11) yields,

$$\frac{\partial A}{\partial t} = \alpha\beta Q^{\beta-1}\frac{\partial Q}{\partial t} \tag{10.16}$$

Therefore, considering Equation (10.9) and Equation (10.16) we can obtain the kinematic wave equation as,

$$\frac{\partial Q}{\partial x} + \alpha\beta Q^{\beta-1}\frac{\partial Q}{\partial t} = q \tag{10.17}$$

This is the kinematic wave equation; it has only one unknown value as Q.

10.5.3. Finite Difference Formulation for Kinematic Wave Equation

The following finite difference approximations can be considered for the derivatives of the function Q in equation (10.17) (Liggett and Woolhisher, 1967; Ligget and Cunge, 1975).

$$\frac{\partial Q}{\partial x} = \frac{Q_{i+1}^{j+1} - Q_i^{j+1}}{\Delta x} \tag{10.18}$$

$$\frac{\partial Q}{\partial t} = \frac{Q_{i+1}^{j+1} - Q_{i+1}^{j}}{\Delta t} \tag{10.19}$$

In addition, the following average value can be written for a function,

$$Q = \frac{Q_{i+1}^{j} + Q_i^{j+1}}{2} \tag{10.20}$$

Let's puts on these finite difference approximations and average function in Equation (10.17) as,

$$\frac{Q_{i+1}^{j+1} - Q_i^{j+1}}{\Delta x} + \alpha\beta \left[\frac{Q_{i+1}^{j} + Q_i^{j+1}}{2}\right]^{\beta-1} \frac{Q_{i+1}^{j+1} - Q_{i+1}^{j}}{\Delta t} = \frac{q_{i+1}^{j+1} + q_{i+1}^{j}}{2} \tag{10.21}$$

Solving Q_{i+1}^{j+1}, the following equation is obtained,

$$Q_{i+1}^{j+1} = \left[\frac{\frac{\Delta t}{\Delta x} Q_i^{j+1} + \alpha\beta\, Q_{i+1}^{j} \left[\frac{Q_{i+1}^{j} + Q_i^{j+1}}{2}\right]^{\beta-1} + \Delta t \frac{q_{i+1}^{j+1} + q_{i+1}^{j}}{2}}{\frac{\Delta t}{\Delta x} + \alpha\beta \left[\frac{Q_{i+1}^{j} + Q_i^{j+1}}{2}\right]^{\beta-1}}\right] \tag{10.22}$$

10.5.4. Formulation for Nonlinear Kinematic Wave Scheme

Again, we can consider the kinematic wave equations, Equation (10.9) and Equation (10.11). The continuity equation, Equation (10.9), can be written in finite difference approximation as (see Figure 10.2)

$$\frac{Q_{i+1}^{j+1} - Q_i^{j+1}}{\Delta x} + \frac{A_{i+1}^{j+1} - A_{i+1}^{j}}{\Delta t} = \frac{q_{i+1}^{j+1} + q_{i+1}^{j}}{2} \tag{10.23}$$

where Q is the flowrate, A is the cross-sectional area and q is the lateral flow.

From Equation (10.11) the relationships between A and Q can be written as

$$A_{i+1}^{j+1} = \alpha\left[Q_{i+1}^{j+1}\right]^{\beta} \tag{10.24}$$

and,

$$A_{i+1}^{j} = \alpha\left[Q_{i+1}^{j}\right]^{\beta} \tag{10.25}$$

Substituting Equation (10.24) and Equation (10.25) into Equation (10.23),

$$\frac{\Delta t}{\Delta x} Q_{i+1}^{j+1} + \alpha\left[Q_{i+1}^{j+1}\right]^{\beta} = \frac{\Delta t}{\Delta x} Q_{i}{}^{j+1} + \alpha\left[Q_{i+1}^{j}\right]^{\beta} + \Delta t \frac{q_{i+1}^{j+1} + q_{i+1}^{j}}{2} \tag{10.26}$$

Using Newton's method, we can write,

$$(Q_{i+1}^{j+1})_{k+1} = (Q_{i+1}^{j+1})_{k} - \frac{f(Q_{i+1}^{j+1})_{k}}{f'(Q_{i+1}^{j+1})_{k}} \tag{10.27}$$

Convergence criterion is defined as

$$\left|f(Q_{i+1}^{j+1})_{k+1}\right| \leq \varepsilon \tag{10.28}$$

Considering (10.26) we can write,

$$f(Q_{i+1}^{j+1}) = \frac{\Delta t}{\Delta x} Q_{i+1}^{j+1} + \alpha(Q_{i+1}^{j+1})^{\beta} - C \tag{10.29}$$

From Equation (10.27) one can write as,

$$f'(Q_{i+1}^{j+1}) = \frac{\Delta t}{\Delta x} + \alpha\beta(Q_{i+1}^{j+1})^{\beta-1} \tag{10.30}$$

Therefore, we obtain

$$C = \frac{\Delta t}{\Delta x} Q_i^{j+1} + \alpha (Q_{i+1}^{j})^{\beta} + \Delta t \, \frac{q_{i+1}^{j+1} + q_{i+1}^{j}}{2} \qquad (10.31)$$

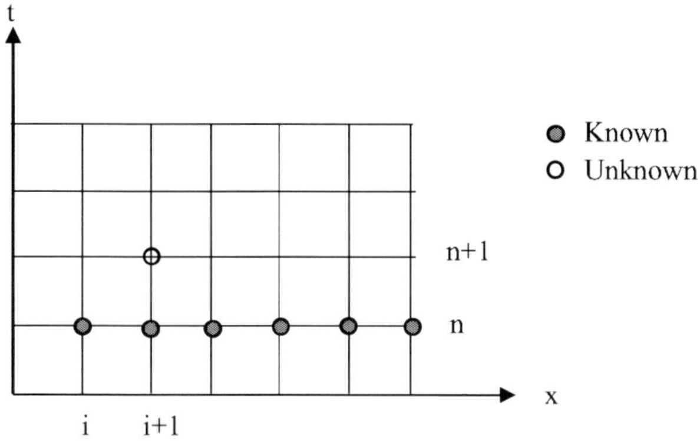

Figure 10.2. Finite difference grid system.

10.6. Applications of Open Channel Flow Modeling

Consider a watershed flow. For the following data:

a. Formulate the kinematic wave equations.
b. Calculate the maximum depth of water at the end of watershed.
c. Calculate the rising depth hydrograph by hand using kinematic model.
d. Calculate the rising depth hydrograph by computer.
e. Plot the rising leg of the hydrograph.

Given data: Plan shape = Rectangular; W = Width = 5 m; L = Length = 22 m; S_0 = Slope = 0.01; i = rainfall intensity = 0.123 m/hour; T= rainfall duration = 840 sec; α = 2.76 and β = 1.70 (kinematic parameters are estimated by optimization).

We calculate the characteristic water depth, h_0.

Solution:

a. Formulation of the kinematic wave equations: Continuity Equation with a lateral inflow, can be written as Equation (10.9),

$$\frac{\partial Q}{\partial x} + \frac{\partial A}{\partial t} = I$$

where Q is the flow rate, A is the cross-sectional area, and I is the lateral inflow.

We assume that the watershed has a rectangular plane shape. For unit wide the equation becomes (A = B×h, B is constant as 5.0 m),

$$\frac{\partial q}{\partial x} + \frac{\partial h}{\partial t} = i$$

in which q is the flow rate per unit wide, h is the water depth, i is lateral inflow per unit wide.

The motion equation for kinematic wave model can be approximated as,

$$q = \alpha h^{\beta}$$

where α, and β are kinematic parameters.

From the above equations, the kinematic wave equation with lateral inflow, i, can be written as:

$$\frac{\partial h}{\partial t} + \alpha\beta h^{\beta-1}\frac{\partial h}{\partial x} = I$$

b. Calculation of maximum water depth: Rational method gives the maximum discharge as:

$$Q_{max} = ciA$$

In which Q_{max} is the maximum discharge, c is the runoff coefficient, i is the rainfall excess, A is the watershed area. For a rectangular watershed A = LB, where l is the watershed length, and B is the watershed wide.

We assume the surface of watershed is impervious. So, c = 1.0. The discharge per unit wide is:

$$q_{max} = ciL = 1.0x(0.123/3600)\times 22 = 7.5167\times 10^{-4}\ m^2/s$$

$$Q_{max} = ciLB = 1.0\times 7.5167\times 10^{-4}\times 5 = 3.7583\times 10^{-3}\ m^3/s = 3.76\ l/s$$

From equation $q = \alpha h^{\beta}$ we can find the maximum water depth at the end of watershed as

$$h_{max} = (q_{max}/\alpha)^{1/\beta} = (7.5167 \times 10^{-4} /2.76)^{1/1.7} = 0.00799825 \text{ m} = 0.8 \text{ cm}$$

c. Solving the problem by hand using finite difference method: Boundary and initial conditions are considered as seen in Figure 10.3:

Upstream boundary condition,

$$h(0, t) \rightarrow h = 0, \qquad \text{for } 0 \leq t < T$$

Initial condition,

$$h(x, 0) \rightarrow h = 0, \qquad \text{for } 0 \leq x < L$$

where T is the rainfall duration, L is the watershed or channel length. Initial and boundary conditions are shown in Figure 10.4.

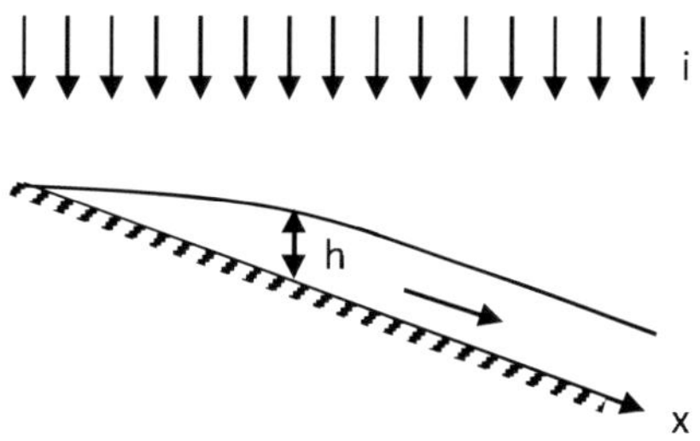

Figure 10.3. Definition sketch of the flow on a watershed.

	j=1	j=2	j=3	j=4	j=5
k=6	0.0				
k=5	0.0				
k=4	0.0				
k=3	0.0				
k=2	0.0				
k=1	0.0	0.0	0.0	0.0	0.0

Figure 10.4. Boundary and initial conditions.

Substituting finite difference approximations (Forward in Time and Central in Space, FTCS) in kinematic wave equation which is derived previously,

$$\frac{h_j^{k+1} - h_j^k}{\Delta t} + \alpha\beta(h_j^k)^{\beta-1}\frac{h_{j+1}^k - h_{j-1}^k}{2\Delta x} = i_j^k$$

This equation is the finite difference form of the governing equation. Solving this equation gives

$$h_j^{k+1} = h_j^k + \Delta t[i_j^k - \alpha\beta(h_j^k)^{\beta-1}(h_{j+1}^k - h_{j-1}^k)/(2\Delta x)]$$

At the downstream grid, the following formula can be used.

$$h_j^{k+1} = h_j^k + \Delta t[i_j^k - \alpha\beta(h_j^k)^{\beta-1}(h_j^k - h_{j-1}^k)/(\Delta x)]$$

Stability criteria can be determined for $h_{max} = h_0 = 0.008$ m:

$$\Delta t/\Delta x \leq 1/(\alpha\beta h_o^{\beta-1}) = 1/(2.76\times1.70\times0.008)^{0.70} = 6.2586$$

or,

$$\Delta t \leq 6.25\Delta x$$

This inequality is the stability criterion for the flow conditions.

For the space interval, $\Delta x = 2$ m, time interval should be as $\Delta t \leq 12.5$ sec.

Time interval is selected as $\Delta t = 5$ sec. We start to calculate the function values as;

For k = 1 and j=2,

$$h_j^{k+1} = h_j^k + \Delta t[i_j^k - \alpha\beta(h_j^k)^{\beta-1} (h_{j+1}^k - h_{j-1}^k)/(2\Delta x)]$$

$$h_2^2 = 0+5\times[(0.123/3600) - (2.76\times1.70\times0)(0 - 0)/(2\times2)] = 1.708\times10^{-4} \text{ m}$$

Similarly for j = 3, 4, .. , and 12,

$h_3^2 = h_4^2 = h_5^2 = h_6^2 = h_7^2 = h_8 = h_9^2 = h_{10}^2 = h_{11}^2 = h_{12}^2 = 1.708\times10^{-4}$ m
For k = 2 and j = 2

$h_2^3 = 0.0001708 + 5[0.123/3600 - 2.76\times1.7\times0.0001708^{0.7})(0.0001798 - 0)/4]$
$= 3.393\times10^{-4}$ m
$h_3^3 = 0.0001708 + 5[0.123/3600 - 2.76\times1.7\times0.0001708^{0.7})(0.0001798- 0.0001708)/4] = 3.4163\times 10^{-4}$ m

Similarly, for j = 3, 4, …, and 11,

$h_4^3 = h_5^3 = h_6^3 = h_7^3 = h_8^3 = h_9^3 = h_{10}^3 = h_{11}^3 = 3.4163\times10^{-4}$ m

At the end of watershed,

$h_{12}^3 = 0.0001708 + 5[0.123/3600 - 2.76\times1.7\times0.0001708^{0.7})(0.0001798 - 0.0001708)/2] = 3.4163\times10^{-4}$ m.

Compare the results by hand with the results by computer.

d. Calculation of rising depth hydrograph at the end of the watershed: In order to calculate the water depths at each points of the mesh system by using computer the following formulas are used

$$h_j^{k+1} = h_j^k + \Delta t[i_j^k - \alpha\beta(h_j^k)^{\beta-1}(h_{j+1}^k - h_{j-1}^k)/(2\Delta x)]$$

At the downstream grid, the following formula can be used.

$$h_j^{k+1} = h_j^k + \Delta t[i_j^k - \alpha\beta(h_j^k)^{\beta-1} (h_j^k - h_{j-1}^k)/(\Delta x)]$$

Calculation is carried out for Δx = 2m, Δt = 5 sec using excel. The computed water depths are shown in Table 10.2. A graph using the calculated rising hydrograph depths versus time at the end of watershed (x = 22 m) is given in Figure 10.5.

e. Calculation of Rising Leg of the Hydrograph: For a = 2.76, b = 1.7 and B = 5 m, using $q = \alpha h^{\beta}$, and Q = qB, one can obtain the water depth values using for interior points as similar manner for problem (d),

$$h_j^{k+1} = h_j^k + \Delta t[i_j^k - \alpha\beta(h_j^k)^{\beta-1}(h_{j+1}^k - h_{j-1}^k)/(2\Delta x)]$$

Table 10.2. Calculated water depths (m)

x = 0	x = 2 m	x = 4 m	x = 6 m	x = 8 m	x = 0 m	x = 12 m	x = 14 m	x = 16 m	x = 18 m	x = 20 m	x = 22 m
0.0000	0.0000	0.0000	0.0000	0.0000	0.0000	0.0000	0.0000	0.0000	0.0000	0.0000	0.0000
0.0000	0.0002	0.0002	0.0002	0.0002	0.0002	0.0002	0.0002	0.0002	0.0002	0.0002	0.0002
0.0000	0.0003	0.0003	0.0003	0.0003	0.0003	0.0003	0.0003	0.0003	0.0003	0.0003	0.0003
0.0000	0.0005	0.0005	0.0005	0.0005	0.0005	0.0005	0.0005	0.0005	0.0005	0.0005	0.0005
0.0000	0.0007	0.0007	0.0007	0.0007	0.0007	0.0007	0.0007	0.0007	0.0007	0.0007	0.0007
0.0000	0.0008	0.0009	0.0009	0.0009	0.0009	0.0009	0.0009	0.0009	0.0009	0.0009	0.0009
0.0000	0.0009	0.0010	0.0010	0.0010	0.0010	0.0010	0.0010	0.0010	0.0010	0.0010	0.0010
0.0000	0.0011	0.0012	0.0012	0.0012	0.0012	0.0012	0.0012	0.0012	0.0012	0.0012	0.0012
0.0000	0.0012	0.0014	0.0014	0.0014	0.0014	0.0014	0.0014	0.0014	0.0014	0.0014	0.0014
0.0000	0.0013	0.0015	0.0015	0.0015	0.0015	0.0015	0.0015	0.0015	0.0015	0.0015	0.0015
0.0000	0.0014	0.0017	0.0017	0.0017	0.0017	0.0017	0.0017	0.0017	0.0017	0.0017	0.0017
0.0000	0.0014	0.0018	0.0019	0.0019	0.0019	0.0019	0.0019	0.0019	0.0019	0.0019	0.0019
0.0000	0.0015	0.0020	0.0020	0.0020	0.0020	0.0020	0.0020	0.0020	0.0020	0.0020	0.0021
0.0000	0.0016	0.0021	0.0022	0.0022	0.0022	0.0022	0.0022	0.0022	0.0022	0.0022	0.0022
0.0000	0.0016	0.0022	0.0024	0.0024	0.0024	0.0024	0.0024	0.0024	0.0024	0.0024	0.0024
0.0000	0.0016	0.0023	0.0025	0.0026	0.0026	0.0026	0.0026	0.0026	0.0026	0.0026	0.0026
0.0000	0.0016	0.0024	0.0027	0.0027	0.0027	0.0027	0.0027	0.0027	0.0027	0.0027	0.0027
0.0000	0.0017	0.0025	0.0028	0.0029	0.0029	0.0029	0.0029	0.0029	0.0029	0.0029	0.0029
0.0000	0.0017	0.0026	0.0029	0.0031	0.0031	0.0031	0.0031	0.0031	0.0031	0.0031	0.0031
0.0000	0.0017	0.0026	0.0031	0.0032	0.0032	0.0032	0.0032	0.0032	0.0032	0.0032	0.0032
0.0000	0.0017	0.0027	0.0032	0.0034	0.0034	0.0034	0.0034	0.0034	0.0034	0.0034	0.0034
0.0000	0.0016	0.0027	0.0033	0.0035	0.0036	0.0036	0.0036	0.0036	0.0036	0.0036	0.0036
0.0000	0.0016	0.0027	0.0034	0.0036	0.0037	0.0038	0.0038	0.0038	0.0038	0.0038	0.0038
0.0000	0.0016	0.0027	0.0034	0.0038	0.0039	0.0039	0.0039	0.0039	0.0039	0.0039	0.0039
0.0000	0.0016	0.0027	0.0035	0.0039	0.0040	0.0041	0.0041	0.0041	0.0041	0.0041	0.0041
0.0000	0.0016	0.0027	0.0035	0.0040	0.0042	0.0043	0.0043	0.0043	0.0043	0.0043	0.0043
0.0000	0.0016	0.0027	0.0035	0.0041	0.0043	0.0044	0.0044	0.0044	0.0044	0.0044	0.0044
0.0000	0.0016	0.0027	0.0036	0.0042	0.0045	0.0046	0.0046	0.0046	0.0046	0.0046	0.0046
0.0000	0.0016	0.0027	0.0036	0.0042	0.0046	0.0047	0.0048	0.0048	0.0048	0.0048	0.0048
0.0000	0.0016	0.0027	0.0036	0.0043	0.0047	0.0049	0.0049	0.0049	0.0050	0.0050	0.0050
0.0000	0.0016	0.0027	0.0036	0.0043	0.0048	0.0050	0.0051	0.0051	0.0051	0.0051	0.0051
0.0000	0.0016	0.0027	0.0035	0.0043	0.0048	0.0051	0.0052	0.0053	0.0053	0.0053	0.0053
0.0000	0.0016	0.0026	0.0035	0.0043	0.0049	0.0052	0.0054	0.0054	0.0055	0.0055	0.0055
0.0000	0.0016	0.0026	0.0035	0.0043	0.0049	0.0053	0.0055	0.0056	0.0056	0.0056	0.0056
0.0000	0.0016	0.0026	0.0035	0.0043	0.0050	0.0054	0.0057	0.0058	0.0058	0.0058	0.0058
0.0000	0.0016	0.0026	0.0035	0.0043	0.0050	0.0055	0.0058	0.0059	0.0060	0.0060	0.0060

x = 0	x = 2 m	x = 4 m	x = 6 m	x = 8 m	x = 0 m	x = 12 m	x = 14 m	x = 16 m	x = 18 m	x = 20 m	x = 22 m
0.0000	0.0016	0.0026	0.0035	0.0043	0.0050	0.0055	0.0059	0.0061	0.0061	0.0061	0.0061
0.0000	0.0016	0.0026	0.0035	0.0042	0.0050	0.0055	0.0060	0.0062	0.0063	0.0063	0.0063
0.0000	0.0016	0.0026	0.0035	0.0042	0.0049	0.0056	0.0060	0.0063	0.0064	0.0065	0.0065
0.0000	0.0016	0.0026	0.0034	0.0042	0.0049	0.0056	0.0061	0.0064	0.0066	0.0066	0.0067
0.0000	0.0016	0.0026	0.0034	0.0042	0.0049	0.0056	0.0061	0.0065	0.0067	0.0068	0.0068
0.0000	0.0016	0.0026	0.0034	0.0042	0.0049	0.0055	0.0061	0.0066	0.0068	0.0069	0.0070
0.0000	0.0016	0.0026	0.0034	0.0042	0.0048	0.0055	0.0061	0.0066	0.0069	0.0071	0.0071
0.0000	0.0016	0.0026	0.0034	0.0042	0.0048	0.0055	0.0061	0.0066	0.0070	0.0072	0.0073
0.0000	0.0016	0.0026	0.0034	0.0042	0.0048	0.0055	0.0061	0.0067	0.0071	0.0073	0.0074
0.0000	0.0016	0.0026	0.0035	0.0041	0.0048	0.0054	0.0061	0.0067	0.0071	0.0074	0.0076
0.0000	0.0016	0.0026	0.0035	0.0042	0.0048	0.0054	0.0060	0.0066	0.0071	0.0075	0.0077
0.0000	0.0016	0.0026	0.0035	0.0042	0.0048	0.0054	0.0060	0.0066	0.0072	0.0076	0.0078
0.0000	0.0016	0.0026	0.0035	0.0042	0.0048	0.0054	0.0060	0.0066	0.0071	0.0076	0.0079
0.0000	0.0016	0.0026	0.0035	0.0042	0.0048	0.0054	0.0059	0.0066	0.0071	0.0077	0.0080
0.0000	0.0016	0.0026	0.0035	0.0042	0.0048	0.0054	0.0059	0.0065	0.0071	0.0077	0.0080
0.0000	0.0016	0.0026	0.0035	0.0042	0.0048	0.0054	0.0059	0.0065	0.0071	0.0077	0.0080
0.0000	0.0016	0.0026	0.0035	0.0042	0.0048	0.0054	0.0059	0.0065	0.0070	0.0076	0.0081
0.0000	0.0016	0.0026	0.0035	0.0042	0.0048	0.0054	0.0059	0.0064	0.0070	0.0076	0.0081

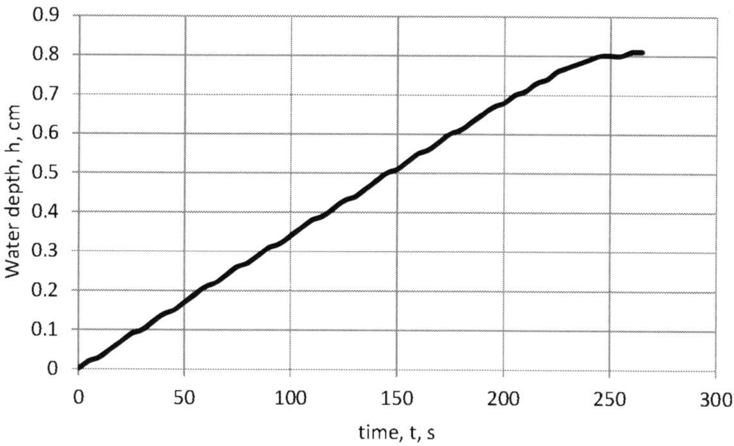

Figure 10.5. Rising depth hydrograph at the end of watershed (x = 22 m).

At the downstream grid, the following formula can be used,

$$h_j^{k+1} = h_j^k + \Delta t[i_j^k - \alpha\beta(h_j^k)^{\beta-1}(h_j^k - h_{j-1}^k)/(\Delta x)]$$

From Table 10.2 we can take h values, and we calculate the corresponding values of discharge per unit wide using $q = \alpha h^\beta$, then we can obtain discharge by $Q = qB$. A graph that shows the amount of water volume per time in the rising leg of the hydrograph is given in Figure 10.6.

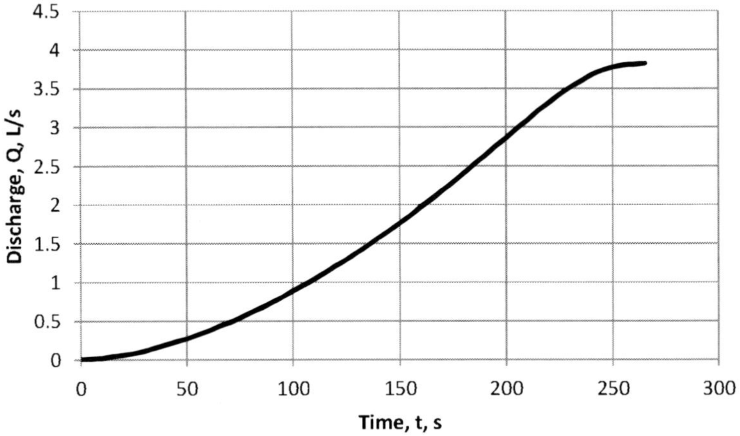

Figure 10.6. Rising Leg of the water discharge hydrograph.

Also, it is clear that the maximum flowrate obtained from rational formula $Q_{max} = 3.76$ L/s is agree with the peak hydrograph value, $Q = 3.82$ L/s.

Chapter 11

Applications of Gradually Varied Flow

11.1. Introduction

In this chapter, gradually varied flows are examined. These are steady and non-uniform flows, characterized by a gradual change in flow depth and velocity with a generally smooth free surface.

Flows with a rapid change in flow depth and velocity are called rapidly varied flows. A change in bottom slope or cross section of the channel or an obstacle in the flow path may cause the uniform flow in the channel to be a gradually or rapidly varied flow, as seen in Figure 11.1.

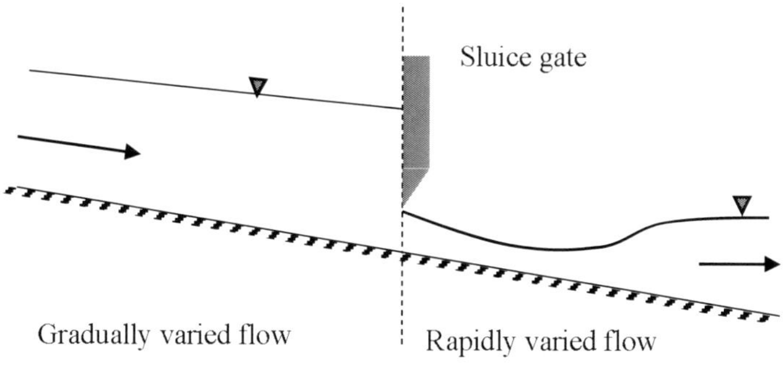

Figure 11.1. Examples for rapidly varied flow, and gradually varied flow.

We are interested in looking at the solution of an ordinary differential equation for the numerical solution of the gradually varied flow. Long channels with flowing water will eventually reach the standard depth, y_n. The uniform flow depth is the name given to this depth. By applying the Manning velocity equation, normal depth can be determined. The water surface profile of a gradually changing flow is a graph of the water depth against the channel length. The water gradually gets to its natural depth in this plot.

A gradually varied flow calculation in a trapezoidal channel involves starting from a known y depth and performing next water depth calculations at small distance intervals. In this method. continuity equation and energy slope equation are used. In calculation, initially normal depth. y_n, critical depth, y_c, and then gradually varied flow profile type are determined.

The downstream component of water weight may not be locally balanced due to disturbances caused by hydraulic structures (weir, sluice, etc.) or variations in channel width, slope, or roughness. The water depth profile is then calculated and plotted. Current properties such as depth, velocity, and Froude number are also displayed in a particular place in this computation. A gradually varied flow profile is also known as water depth profile, backwater calculation, and non-uniform flow calculation. This calculation is for steady state flow.

In normal flow, the downstream component of the water weight is balanced by bed friction. As a result, water depth y and current velocity V are constant, and the energy slope line is parallel to both the water surface and the channel bottom. In other words, the friction slope, S_f, is the same as the geometric slope of the channel, S_o.

Briefly, in gradually varied flow, the friction slope, Sf, and the bottom slope, S_o, will differ from each other, and the water depth, y, and flow velocity, V, will change along the channel. An expression will be derived for the equation of the gradually varied flow for dx/dt and this expression can be used to predict water depth variation along the channel.

11.2. Derivation of the Gradually-Varied-Flow Equation

In order to derive the energy equation for gradually varied flow, considering Figure 11.2, the total head can be written for two sections as,

$$H = z_1 + y_1 + \frac{V_1^2}{2g} = z_2 + y_2 + \frac{V_2^2}{2g} + h_L \qquad (11.1)$$

in which g is the acceleration due to gravity, 9.8066 m/s^2,

V is the average velocity of water [m/s],

H is the total head [m],

y is the water depth [m], h_L is the friction loss [m],

z is the channel bottom elevation from datum [m].

In this part, we accept small slope and also accept that there is no difference between the vertical depth, y, that perpendicular to the channel bed, ycosθ.

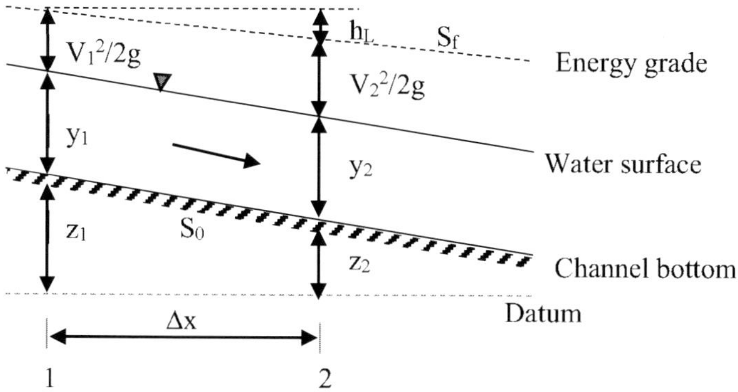

Figure 11.2. Water surface profile for an open channel flow.

Considering Figure 11.2, for total head we can write from Bernoulli equation as,

$$H = z + E \tag{11.2}$$

where H is the total head, E is the specific energy, and z is the height above datum level.

For specific energy we can write,

$$E = y + \frac{V^2}{2g} \tag{11.3}$$

In frictionless flows, H is constant, that is. the energy slope line is horizontal. In reality, due to bed friction, H decreases when large distances are exceeded and the energy slope line slopes downward.

Differentiating Equation (11.2) yields,

$$\frac{dH}{dx} = \frac{dz}{dx} + \frac{dE}{dx} \tag{11.4}$$

where x is the horizontal distance.

We can define,

$$\frac{dH}{dx} = -S_f \tag{11.5}$$

$$\frac{dz}{dx} = -S_0 \tag{11.6}$$

where S_f is the friction slope, S_0 is the actual geometric slope.

From Equation (11.2) we can write

$$\frac{dE}{dx} = S_0 - S_f \tag{11.7}$$

As a result, the specific energy only varies when the friction gradient differs from the geometrical gradient. If not, we will experience normal flow, constant water depth, and specific energy.

Equation (11.5), and Equation (11.7) are two different forms of the gradually varied flow equation. A third one, most common form rewrites dE/dx, in terms of the rate of change of water depth, dy /dx, as Equation (11.3).

Considering the continuity equation, we can write

$$Q = VA \tag{11.8}$$

Substituting Equation (11.8) into Equation (11.3),

$$E = y + \frac{Q^2}{2gA^2} \tag{11.9}$$

where, E is the specific energy head [m],

y is the water depth [m],

A is the channel cross-sectional area [m^2], Q is the discharge [m^3/s],

g is the gravitational acceleration [m/s^2],

Vis the average velocity of water [m/s]. The following geometric relationships are always valid for trapezoidal channels, using Q = VA, as seen in Figure 11.3.

$$T = b + y\,(z_1 + z_2) \tag{11.10}$$

$$A = y(b + T)/2 \tag{11.11}$$

$$P = b + y(1 + z_1^2)^{0.5} + y(1 + z_2^2)^{0.5} \tag{11.12}$$

Setting R =A/P in Equation (11.11) and Equation (11.12)

$$R = \frac{y(b + T)/2}{b + y(1 + z_1^2)^{0.5} + y(1 + z_2^2)^{0.5}} \tag{11.13}$$

in which, A is the cross-sectional area [m^2],

b is the channel bottom width [m],

P is the channel wetted perimeter [m],

R is the hydraulic radius [m],

T is the top width of the channel [m],

y is the water depth [m],

z_1 is the one channel side slope (horizontal to vertical ratio) [m/m],

z_2 is the other channel side slope (horizontal to vertical ratio) [m/m].

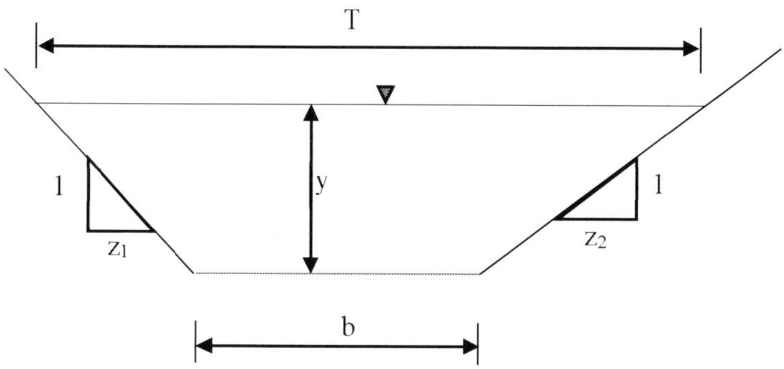

Figure 11.3. A cross section of the trapezoidal channel.

Differentiating Equation (11.9) with respect to streamwise distance x:

$$\frac{dE}{dx} = \frac{dy}{dx} - \frac{Q^2}{gA^3}\frac{dA}{dx} \tag{11.14}$$

If T is the width of the channel at the water surface (see Figure 11.3), we can write:

$$dA = T dy \tag{11.15}$$

Substituting Equation (11.15) into Equation (11.14) and rearranging,

$$\frac{dE}{dx} = \frac{dy}{dx}\left[1 - \frac{Q^2 T}{gA^3}\right] \tag{11.16}$$

Substituting Equation (11.7) into Equation (11.16),

$$\frac{dy}{dx} = \frac{S_0 - S_f}{1 - Q^2 T/(gA^3)} \tag{11.17}$$

This is the gradually varied flow equation, where Q is the discharge (m^3/s), T is the top width of the water surface (m), and A is the water cross sectional area, (m^2), S_0 is the bottom slope of the channel, S_f is the friction or energy slope, and dy/dx is the variation in flow depth with respect to distance x.

We can write this equation in another form. Using V = Q/A, gives

$$\frac{dy}{dx} = \frac{S_0 - S_f}{1 - V^2 T/(gA)} \quad (11.18)$$

For a trapezoidal channel Froude number is given by

$$F = \frac{V}{(g\frac{A}{T})^{0.5}} \quad (11.19)$$

A is the cross-sectional area [m^2],
T is the top width of the channel [m],
F is the Froude number [dimensionless],
V is the average velocity of water [m/s],
g is the gravitational acceleration [m/s^2].

Putting Equation (11.19) into Equation (11.18) we obtain the other form of gradually varied flow as,

$$\frac{dy}{dx} = \frac{S_0 - S_f}{1 - F^2} \quad (11.20)$$

in which S_0 is the bottom slope of the channel, S_f is the friction or energy slope, F is the Froude number, dy/dx is the variation flow depth with respect to distance x.

11.3. Determination of the Friction Slope

Since the velocity and hence the flow changes only gradually with distance, the friction mainly changes with the local cross-section velocity, V. The local friction slope S_f can then be evaluated with the assumption of "semi-uniform flow" where the same depth has the same energy loss rate as in normal flow (Shieh, 2009).

Using the Manning equation, we can write:

$$V = \frac{1}{n} R^{2/3} S_f^{1/2} \tag{11.21}$$

in which R is the hydraulic Radius, n is the Manning friction coefficient, S_f is the friction slope. Manning friction coefficient can be obtained from Table 11.1.

Inverting for the friction slope, gives

$$S_f = \frac{n^2 V^2}{R^{4/3}} = \frac{n^2 Q^2}{R^{4/3} A^2} \tag{11.22}$$

The cross-sectional area of water, A, and the hydraulic radius, R, would be written in terms of depth, y. In general, as the water depth increases, the velocity and becomes smaller. the deeper the flow, the smaller the velocity and friction losses.

Table 11.1. Friction coefficients, n, for the Manning equation for channels

Channel Material	Coefficient, n	Channel Material	Coefficient, n
Natural Streams		*Excavated Earth Channels*	
Clean and Straight	0.030	Clean	0.022
Major Rivers	0.035	Gravelly	0.025
Sluggish with Deep Pools	0.040	Weedy	0.030
		Stony, Cobbles	0.035
Metals		Floodplains	
Brass	0.011	Pasture, Farmland	0.035
Cast Iron	0.013	Brush, Light	0.050
Steel, Smooth	0.012	Brush, Heavy	0.075
Steel, Corrugated	0.022	Trees	0.150
Non-Metals			
Glass	0.010	Concrete, Finished	0.012
Clay Tile	0.014	Concrete, Unfinished	0.014
Brickwork	0.015	Gravel	0.029
Asphalt	0.016	Earth	0.025
Masonry	0.025	Wood, Planed	0.012
		Wood, Unplaned	0.013

11.4. Classification of Free Surface Profiles

If Froude Number, F, equals 1 in a channel flow, the resulting depth is called critical depth and it is usually denoted by y_c. In a wide and rectangular cross-section channel, the critical depth is given by $y_c = (q^2/g)^{1/3}$. On the other hand, a depth occurs for a certain flow and slope in a uniform flow, and its water depth is known as normal depth, abbreviated y_n.

If the normal depth is less than the critical depth, the channel slope is called as steep (S). In this case, normal flow is known as supercritical flow. If the normal depth is greater than the critical depth, the channel slope is referred as mild (M), and in this case the normal flow is known as subcritical flow. In fact, since the depths depend on the rate of flow in the channel, the steep or mild slope occurs according to the rate of flow.

In general, after determining the critical depth, y_c, and normal depth, y_n, the real flow depth, y, is determined from the right hand side of the gradually varied flow equation given in Equation (11.20) as, When the right hand side of Equation (11.20) is positive, the depth should be increasing or if the right hand side of this equation is negative, the depth should be decreasing with distance.

If $S_f = S_0$, the flow will be normal flow;
If $S_0 - S_f > 0$, the water depth, y, should be greater than normal depth;
If $1 - F^2 > 0$, the water depth, y, should be greater than critical depth.

In classification of the water surface slope, we use some symbols such as S_1, M_2 etc. In this classification, we can use the first character for slopes as S, C, M, H, A for Steep, Critical, Mild, Horizontal, Adverse, respectively; and the second character as 1, 2, 3 depending on where water depth, y lies with respect to y_c and y_n.

Depending on the zone and slope of the channel slope, the water surface profiles are classified into 13 types as seen in Table 11.2 (Dubin, 1999).

Table 11.2. Types of water surface profile

Slope curves	Types of water surface profiles		
	Zone 1	Zone 2	Zone 3
Mild slope curves ($y_n>y_c$)	M_1 ($y>y_n>y_c$)	M_2 ($y_n>y>y_c$)	M_3 ($y_n>y_c>y$)
Steep slope curves ($y_c>y_n$)	S_1 ($y>y_c>y_n$)	S_2 ($y_c>y>y_n$)	S_3 ($y_c>y_n>y$)
Critical slope curves ($y_c=y_n$)	C_1 ($y>y_n=y_c$)	C_1	C_3 ($y_n=y_c>y$)
Horizontal slope curves ($S_0=0$)	(H_1 is not possible)	H_2 ($y>y_c$)	H_3 ($y_c>y$)
Adverse slope curves ($S_0<0$)	(A_1 is not possible)	A_2 ($y>y_c$)	A_3 ($y_c>y$)

11.5. Formulation of the Gradually Varied Flow Equation

Since the gradually varied flow equation can be solved rarely analytically, the solution of this equation is usually numerical. A series of xi and depth yi's at that point are calculated with the operations performed. If necessary, interpolation can be used to calculate the depth values at the intermediate points.

Total energy can be expressed using the Bernoulli equation as $H = Z + V^2/(2g)$. Equation (11.5) is the friction slope obtained by integrating H.

Equation (11.7), $E = y + V^2/(2g)$, can be written for specific energy.

Equation (11.20) can be used to determine the water depth, y.

The friction slope (11.22) is calculated from Manning's equation as in Equation (11.5), (11.7), or Another term for the Froude number is given in Equation (11.19).

In practice, the calculations should start from a control point and continue from there. If the flow is subcritical flow, calculation should proceed forward in x, and this is called downstream control flow. If the flow is supercritical flow, the calculation should proceed backward, and this is called upstream control flow.

The solution methods can be classified as:

Standard-step methods: This method solve for depth y at specified distance intervals Δx.

Direct-step methods: this method solve the equation for distance x at specified depth intervals Δy.

11.5.1. Total-Head Form of the Gradually Varied Flow Equation

The total energy grade slope can be written as in Equation (11.5)

$$\frac{dH}{dx} = -S_f$$

This can be solved using a standard-step method. The equation can be discretized as,

$$\frac{H_{i+1} - H_i}{\Delta x} = -\frac{S_{f.i} - S_{f.i+1}}{2} \tag{11.23}$$

In this method, starting from the control point y_o, water depths y_1, y_2, y_3... are calculated. Since H, total head. H and friction slope, S_f, are functions of y, it is necessary to perform iterative calculations at each calculation step. After first iteration, the iterations will continue until the right hand side and the left hand side of the Equation (11.23) are equal. Although this method is good, it is difficult to make these iterations manually or with a calculator. Therefore, a computer program is required in the calculation (Santiago-Collazo and Araya, 2014).

11.5.2. Specific-Energy Form of the Gradually Varied Flow Equation

The slope of the specific energy is given in Equation (11.7) as:

$$dE/dx = S_0 - S_f$$

where $E = y + V^2/(2g)$.

For a given depth intervals, the displacement can be obtained from Equation (11.31). In this method for values of y, the increments in specific energy, E, can be calculated directly. For this reason, this method is called direct-step method.

Inverting Equation (11.7) yields:

$$\frac{dx}{dE} = \frac{1}{S_o - S_f} \tag{11.24}$$

Equation (11.24) is then discretized as $dx/dE \approx \Delta x/\Delta E$, and rearranged for distance intervals as:

$$\Delta x = \frac{\Delta E}{(S_o - S_f)_{av}} \tag{11.25}$$

where

$$\Delta E = E_{i+1} - E_i \tag{11.26}$$

In order to estimate the average slope difference, there are some alternatives.

11.5.3. Depth Form of the Gradually Varied Flow Equation

We can consider Equation (11.20) as,

$$\frac{dy}{dx} = \frac{S_o - S_f}{1 - F^2}$$

This equation can be solved using a direct-step method.
Inverting Equation (11.20), makes y as independent variable as;

$$\frac{dx}{dy} = \frac{1 - F^2}{S_o - S_f} \tag{11.27}$$

Equation (11.27) can be discretized as $dx/dy \approx \Delta x/\Delta y$, and rearranged for distance intervals as:

$$\Delta x = \frac{dx}{dy}\Delta y \tag{11.28}$$

In Equation (11.28), the bracketed term on the right hand side can be calculated as the average of values at the beginning and end of an interval using evaluation at the interval mid-depth as:

$$y_{mid} = \frac{1}{2}(y_i + y_{i+1}) \tag{11.29}$$

The finite difference form for Equation (11.20) can be written;

$$y_{i+1} - y_i = \Delta x \frac{S_o - S_f}{1 - F^2} \tag{11.30}$$

or

$$\Delta x = (y_{i+1} - y_i)\frac{1 - F^2}{S_o - S_f} \tag{11.31}$$

In order to find Δx, Equation (11.31) can be solved easily, if we know the value of y_i by selecting any value of y_{i+1}.

11.6. Applications of Gradually Varied Flow

Consider a trapezoidal canal, for the following data:

a. Compute the normal depth in the channel.
b. Compute the critical depth in the channel.
c. Determine the type of water surface profile.
d. Compute the backwater profile upstream of the dam using excel.
e. Predict the distance upstream of the dam at where water depth is 4.0 m.
f. Plot the graph of backwater profile.

Given data: Discharge: $Q = 90\ m^3/s$; Bottom wide: b= 15 m; Right sidewall slope: $z_1 = 2$, (1V:2H); Left Sidewall slope: $z_2 = 3$, (1V:3H); Manning friction coefficient: n = 0.035; Bottom slope: So = 0.001.

Solution: Consider Figure 11.3.

a. Computation of the Normal Depth: In order to compute normal depth y_n, consider Figure 11.3. In this case, the Manning Equation can be used for obtaining the normal depth,

$$Q = VA = \frac{1}{n}(A/P)^{2/3}(S_o)^{1/2}A$$

in which Q is the discharge $[m^3/s]$,

A is the cross-sectional area $[m^2]$,

n is the Manning friction coefficient (dimensionless),

P is the channel wetted perimeter [m],

R is the hydraulic Radius = P/A, [m], V is the average velocity of water [m/s], So is the bottom slope [m/m].

We select any water depth, y, say y = 1. For y =1, we obtain from Equations (11.10), (11.11). (11.12), and (11.13):

Top width of the channel, $T = b + y\ (z_1 + z_2) = 15 + 1x(2 + 3) = 20$ m.

Cross sectional area, $A = y(b + T)/2 = 1(15 + 20)/2 = 17.500\ m^2$.

Wetted perimeter, $P = b + y(1+ z_1^2)^{0.5} + y(1+ z_2^2)^{0.5} = 15 + 1(1+4)^{0.5} + 1(1 + 9)^{0.5} = 20.398$ m.

Hydraulic Radius, R = A/P = 17.5/20.398 = 0.858 m.

Discharge, $Q = (1/n)\,(R)^{2/3}(S_o)^{1/2}A = ((1/0.035)(0.858)^{2/3}(0.001)^{1/2}(17.5)$ $= 14.276\ m^3/s$.

The results of these calculations are shown in Table 11. 3 on the first raw.

We continue the normal depth calculation with other values of y untill obtaining discharge equals to 90 m^3/s. These calculations are shown for other y values in Table 11.3.

Table 11.3. Normal depth calculation using excel

Water Depth	Wetted Area	Perimeter	Hydraulic Radius			Discharge
y	A	P	R	$R^{2/3}$	$So^{0.5}$	Q
m	m^2	m	m		-	m^3/s
1.000	17.500	20.398	0.858	0.903	0.032	14.276
2.000	40.000	25.797	1.551	1.340	0.032	48.416
3.000	67.500	31.195	2.164	1.673	0.032	102.027
2.800	61.600	30.115	2.045	1.611	0.032	89.682
2.806	61.774	30.148	2.049	1.613	0.032	**90.041**

The normal depth is calculated as, $y_n = 2.806$ m.

b. Computation of the Critical Depth: The Froude number should be 1 at a critical depth. An example of a prismoidal channel's Froude number is:

$$F = V/[(g(A/T)]^{0.5}$$

where

$$T = b + y\,(z_1 + z_2)$$

$$A = y(b + T)/2$$

in which F is the Froude number [dimensionless],

A is the cross-sectional area [m^2],

g is the gravitational acceleration [m/s^2],

b is the channel bottom width [m],

T is the top width of the channel [m],

V is the average velocity of water [m/s],

y is the water depth [m],

z_1 is the one channel side slope (horizontal to vertical ratio) [m/m],

z_2 is the other channel side slope (horizontal to vertical ratio) [m/m].

In order to calculate the critical depth, at the begining we select a depth say y = 1 m.

For y =1, we can calculate A =17.5 m^2.

Top width of the channel, T, is calculated 20 m. for y =1 m.

From continuity equation, we calculate velocity as V= Q/A = 90/17.5 = 5.143 m/s

Froude number for y =1.0 m, as $F = V/[(g(A/T)]^{0.5} = 5.143/[9.81(17.5/20)]^{0.5} = 1.755$.

The results of these calculations are shown in Table 11.4 on the first raw.

We continue to the critical depth calculation with other values of y untill obtaining Froude number equals to 1. These calculations are shown for other y values in Table 11.4.

Table 11.4. The critical depth calculations using excel

Water Depth	Wetted Area	Flow Velocity	Froude Number
y	A	V	F
m	m^2	m/s	-
1.000	17.5	5.143	1.755359
1.500	28.125	3.200	0.91382
1.400	25.9	3.475	1.022515
1.420	26.341	3.417	**0.999208**

We calculate the critical depth as, $y_c = 1.42$ m.

c. Determination of the Profile Type: Since $y_c = 1.42 < y_n = 2.806 < y = 5.0$, from Table 11.2 we can see that type of water surface profile is type M_1. This situation corresponds to a mild slope ($y_c < y_n$).

d. Backwater Calculation Using Excel: As seen from Equation (11.3), the specific energy equation can be written as,

$$E = y + \frac{V^2}{2g}$$

where α is the energy correction coefficient. On the other hand, we can write from Equation (11.7) as,

$$\frac{dE}{dx} = S_0 - S_f$$

Inverting the above equation, we can write as Equation (11.24):

$$\frac{dx}{dE} = \frac{1}{S_o - S_f}$$

This equation is discretized as $dx/dE \approx \Delta x/\Delta E$ and rearranged for distance intervals as Equation (11.25):

$$\Delta x = \frac{\Delta E}{(S_o - S_f)_{av}}$$

where

$$\Delta E = E_{i+1} - E_{i,}$$

$$E_i = y_i + V_i^2/(2g)$$

Calculation starts with known value of 5 m as water depth, y_i. We calculated the wetted area, wetted perimeter and some other parameters for y = 5 m. The calculated values are: A = 137.5 m^2, P = 41.992 m, R = 3.274 m, $R^{4/3}$ = 4.862, V = 0.655 m/s, $V^2/2g$ = 0.024 m, y +$V^2/2g$ = 5.024 m, S_f = 0.00011, and the normal depth level (NDL) elevation = 67.806 m. Then we show the calculated values on the first row of Table 11.5.

Then, we select 0.1 m as an interval, so we obtain y_{i+1} = 4.9 m. Similarly, we continue to compute all other parameters as shown in Table 11.5.

e. Determination of Distance: For y = 4.0 m, we can determine the corresponding distance by reading from Table 11.5 as x = 1182.788 m.

f. Graph of the Water Surface Profile: The calculated backwater profile, normal depth, and critical depth are shown in in Table 11.6, and Figure 11.4.

Table 11.5. Calculation of backwater profile

Channel Bottom Elevation	Water Surface Elevation	Water Depth y	Wetted Area A	Velocity V	$V^2/2g$	$y + V^2/2g$	ΔE	S_f	Avg S_f	S_0 - Avg S_f	Δx	x	Normal Depth Level
m	m	m	m^2	m/s	m	m	m	10^{-4}	10^{-4}	10^{-4}	m	m	m
65.000	70.000	5	137.5	0.655	0.024	5.024	…..	1.1	-	-	-	0	67.806
65.111	70.011	4.9	133.525	0.674	0.025	4.925	0.099	1.2	1.1	8.9	111.036	111.036	67.917
65.334	70.134	4.8	129.6	0.694	0.027	4.827	0.098	1.3	1.2	8.8	112.109	223.145	68.14
65.671	70.371	4.7	125.725	0.716	0.029	4.729	0.098	1.4	1.3	8.7	113.324	336.469	68.477
66.122	70.722	4.6	121.9	0.738	0.031	4.631	0.098	1.5	1.4	8.6	114.705	451.174	68.928
66.689	71.189	4.5	118.125	0.762	0.033	4.533	0.098	1.6	1.6	8.4	116.282	567.456	69.495
67.375	71.775	4.4	114.4	0.787	0.035	4.435	0.098	1.8	1.7	8.3	118.095	685.551	70.181
68.181	72.481	4.3	110.725	0.813	0.037	4.337	0.098	2.0	1.9	8.1	120.192	805.743	70.987
69.109	73.309	4.2	107.1	0.84	0.04	4.24	0.097	2.1	2.1	7.9	122.634	928.377	71.915
70.163	74.263	4.1	103.525	0.869	0.042	4.142	0.097	2.4	2.3	7.7	125.504	1053.88	72.969
71.346	75.346	4	100	0.9	0.045	4.045	0.097	2.6	2.5	7.5	128.907	1182.79	74.152
72.661	76.561	3.9	96.525	0.932	0.049	3.949	0.097	2.9	2.7	7.3	132.99	1315.78	75.467
74.115	77.915	3.8	93.1	0.967	0.052	3.852	0.096	3.2	3.0	7.0	137.953	1453.73	76.921
75.865	79.465	3.6	86.4	1.042	0.061	3.661	0.192	3.9	3.5	6.5	296.198	1749.93	78.671
77.777	81.277	3.5	83.125	1.083	0.066	3.566	0.095	4.3	4.1	5.9	161.76	1911.69	80.583
79.863	83.263	3.4	79.9	1.126	0.071	3.471	0.095	4.8	4.6	5.4	175.018	2086.71	82.669
82.144	85.444	3.3	76.725	1.173	0.077	3.377	0.094	5.4	5.1	4.9	193.41	2280.12	84.95
84.644	87.844	3.2	73.6	1.223	0.084	3.284	0.093	6.1	5.8	4.2	220.436	2500.55	87.45
87.408	90.508	3.1	70.525	1.276	0.091	3.191	0.093	6.9	6.5	3.5	263.677	2764.23	90.214
90.516	93.516	3	67.5	1.333	0.1	3.1	0.092	7.8	7.3	2.7	343.22	3107.45	93.322
94.159	97.059	2.9	64.525	1.395	0.109	3.009	0.091	8.8	8.3	1.7	535.481	3642.93	96.965
98.646	101.476	2.83	62.472	1.441	0.116	2.946	0.063	10.0	10.0	0	843.864	4486.8	98.646

Table 11.6. Calculated values of backwater depth, normal depth and critical depth

Distance from structure	Bottom Level	Backwater Level	Normal Depth Level	Critical Depth Level
-x	z	E_b	E_n	E_c
m	m	m	m	m
0.00	0.00	5.00	2.81	1.42
-111.04	0.11	5.01	2.92	1.53
-223.15	0.22	5.02	3.03	1.64
-336.47	0.34	5.04	3.14	1.76
-451.17	0.45	5.05	3.26	1.87
-567.46	0.57	5.07	3.37	1.99
-685.55	0.69	5.09	3.49	2.11
-805.74	0.81	5.11	3.61	2.23
-928.38	0.93	5.13	3.73	2.35
-1053.88	1.05	5.15	3.86	2.47
-1182.79	1.18	5.18	3.99	2.60
-1315.78	1.32	5.22	4.12	2.74
-1453.73	1.45	5.25	4.26	2.87
-1749.93	1.75	5.35	4.56	3.17
-1911.69	1.91	5.41	4.72	3.33
-2086.71	2.09	5.49	4.89	3.51
-2280.12	2.28	5.58	5.09	3.70
-2500.55	2.50	5.70	5.31	3.92
-2764.23	2.76	5.86	5.57	4.18
-3107.45	3.11	6.11	5.91	4.53
-3642.93	3.64	6.54	6.45	5.06
-4486.80	4.49	7.32	7.29	5.91

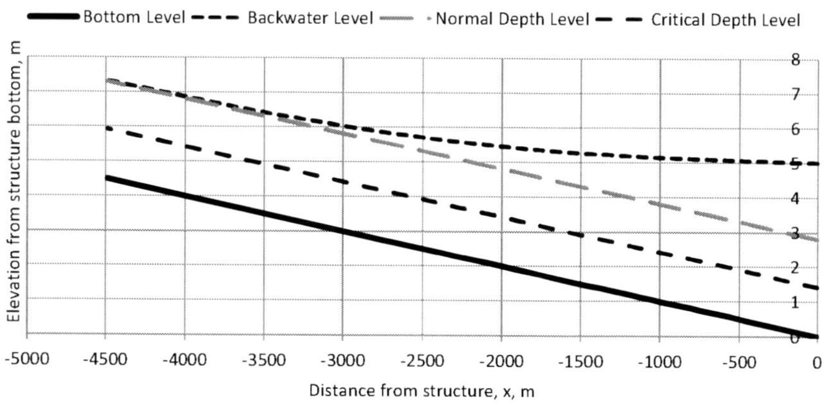

Figure 11.4. Backwater profile, normal depth and critical depth.

Chapter 12

Applications by Method of Characteristics

12.1. Introduction

The method of characteristics was derived from a study of the French mathematician Gaspard Monge (1746-1818) and was used for the first time in the graphical solution of a partial differential equation by Belgian engineer Junius Massau (Massau, 1900). Today, this method is recognized as the most sensitive and reliable of all numerical integral approximations.

The method of characteristics is widely used in solving partial differential equations of problems encountered in science and engineering branches. Partial differential equations are well suited in science and engineering for modeling physical systems and processes. Analytical solution of partial differential equations is complex and difficult. Moreover, these solutions are impractical. Although many approximate solutions are used for the solution of partial differential equations, the method of characteristics is both suitable and practical in computer solutions. It provides more reliable and safe design especially in engineering fields.

This method is a technique used in the solution of partial differential equations. For example, although this applies to first-order equations, more generally it is valid for any hyperbolic partial differential equation. In fact, the method converts a partial differential equation into two ordinary differential equations. The solutions can only be found on the curve, whose initial condition is known and is called the characteristic curve.

In this Chapter, the basic principles of the method of characteristics will be explained step by step for a few engineering studies. In addition, some examples of solutions related to the applications of the method will be presented.

12.2. Method of Characteristics in Transient Flow

Method of characteristics has been used commonly in transient flows which can be expressed in terms of partial differential equations. To solve this partial differential equations, most of the time it is convenient to use method of characteristics which has an ability to convert partial differential equations into ordinary differential equations. Then, these ordinary differential equations are solved through characteristic lines using finite difference approximations.

Transient flows occur in the pipelines, mostly by closing valves as can be seen in Figure 12.1, opening and closing pumps, and other factors that cause the current to accelerate or slow down. Newton was the first to begin researching the analysis of transient flow. Later, the method of characteristics was found for the solution of partial differential equations and this method was completed by some other researchers in the 1950s (Salmanzadeh and Torfi, 2011).

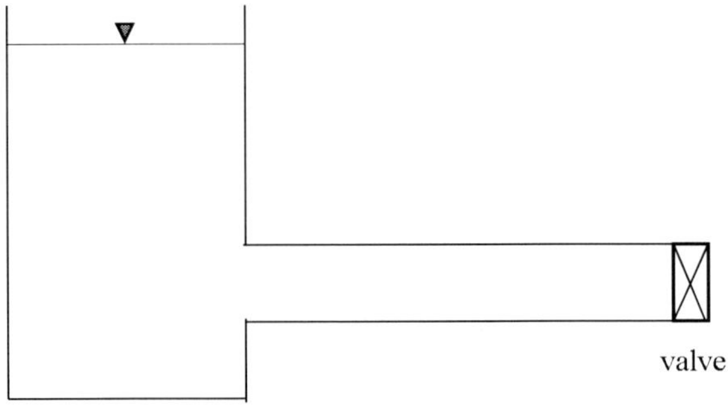

Figure 12.1. A typical transient flow in a pipe at the end of a reservoir.

Transient flow in pipelines mainly is a result of operations or failure of hydraulic appurtenances located on the pipe lines such as closing of the valve, a sudden stopping or starting of the pump. This condition is commonly known as water hammer or hydraulic shock. The pressure waves occurred as a result of these events in pipelines travel at a constant celerity. For computing hydraulic conditions in pipe lines, there are two methods: 1. Interpolation method, 2. Combination of characteristics method and the implicit finite difference method.

Both continuity and momentum equation are considered as basic equations in transient flows. The equations governing hydraulic transients are a pair of coupled, non-linear, first-order partial differential equations that are given below.

$$V\frac{\partial H}{\partial x} + \frac{\partial H}{\partial t} + \frac{c^2}{g}\frac{\partial V}{\partial x} = 0 \tag{12.1}$$

$$\frac{\partial V}{\partial t} + V\frac{\partial V}{\partial x} + g\frac{\partial H}{\partial x} + \frac{fV|V|}{2D} = 0 \quad (12.2)$$

where c = Pressure wave speed, m/s; g = Acceleration of gravity, 9.81 m/s; H = Hydraulic grade line elevation, m; V = Velocity, m/s; D = Pipe diameter, m; f is the Darcy – Weisbach coefficient; x and t are spatial and time dimension, respectively. These equations can be simplified by assuming that the advective terms $V\partial V/\partial x$ and $V\partial H/\partial x$, are very small compared to other terms. Thus, these terms can be negligible and the resulting equations are:

$$\frac{\partial H}{\partial t} + \frac{c^2}{g}\frac{\partial V}{\partial x} = 0 \quad (12.3)$$

$$\frac{\partial V}{\partial t} + g\frac{\partial H}{\partial x} + \frac{fV|V|}{2D} = 0 \quad (12.4)$$

We are required to find the values of H(x,t) and V(x,t) in a domain defined by $0 < x < L$, and $0 < t < T$, where L is the length of the pipe, and T is the maximum duration for the time domain.

In this problem, the characteristic lines can be defined in the x-t plane by dx/dt = u(x,t). It is known that along these characteristic lines function of H and V is conserved in time. Since the continuity and momentum equations given above consist of partial derivatives of H and V with respect to x and t, it is possible to obtain total derivatives of H and V from the combination of these equations.

It is convenient to add continuity equation multiplied by a scale factor λ to the momentum equation.

$$\lambda\left(\frac{\partial H}{\partial t} + \frac{c^2}{g}\frac{\partial V}{\partial x}\right) + \frac{\partial V}{\partial t} + g\frac{\partial H}{\partial x} + \frac{fV|V|}{2D} = 0 \quad (12.5)$$

$$\lambda\left(\frac{\partial H}{\partial t} + \frac{g}{\lambda}\frac{\partial H}{\partial x}\right) + \left(\frac{\partial V}{\partial t} + \frac{\lambda c^2}{g}\frac{\partial V}{\partial x}\right) + \frac{fV|V|}{2D} = 0 \quad (12.6)$$

It is clear from the above expression that the terms in the parentheses correspond to dH/dt and dV/dt, respectively. So, $dx/dt = g/\lambda$ for dH/dt and $dx/dt = \lambda c^2/g$ for dV/dt. Since we are dealing with same physical event, expressions for dx/dt must be equal to each other.

$$\frac{g}{\lambda} = \frac{\lambda c^2}{g}$$

or

$$\lambda = \pm \frac{g}{c}$$

Therefore, the characteristic lines are written as:

$$\frac{dx}{dt} = \pm c$$

Since pressure wave speed, c, calculated is a constant value for the pipeline, the characteristic lines are linear lines with slopes +c (denoted as a C+ line) and –c (denotes as a C- line). After the necessary arrangement the final equation reduces to the combined equation given above:

$$\frac{dH}{dt} \pm \frac{c}{g}\frac{dV}{dt} \pm \frac{cf}{2gD}V|V| = 0 \tag{12.7}$$

If we use Q = VA instead of V, the characteristic equations can be re-written as

$$\frac{dH}{dt} + \frac{c}{gA}\frac{dQ}{dt} + \frac{cf}{2gDA^2}Q|Q| = 0, \text{ along C+: } \frac{dx}{dt} = +c$$

$$\frac{dH}{dt} - \frac{c}{gA}\frac{dQ}{dt} - \frac{cf}{2gDA^2}Q|Q| = 0, \text{ along C-: } \frac{dx}{dt} = -c$$

Consider the sub-grid in the solution domain shown in Figure 12.2.

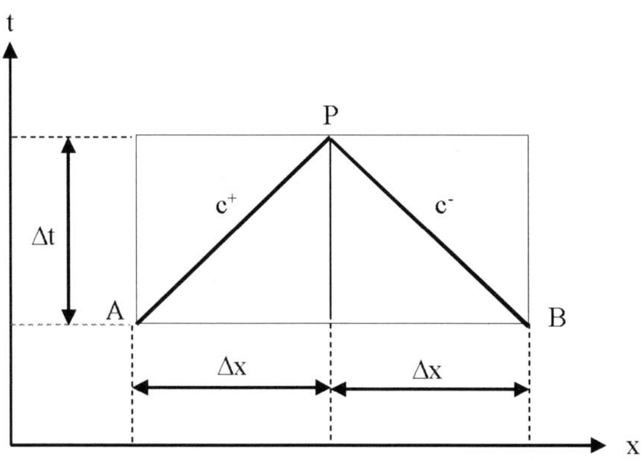

Figure 12.2. Definition of characteristic lines.

Taking the first solution along c+ and substituting M = c/(gA), N = f/(2gdA^2), and c = dx/dt, it can be found:

$$\frac{dH}{dt} + M\frac{dQ}{dt} + NQ|Q|\frac{dx}{dt} = 0 \quad (12.8)$$

Eliminating dt and integrating along the C+ line, i.e., from point (A) to point (P) generates:

$$\int_A^P dH + M\int_A^P dQ + NQ|Q|\int_A^P dx = 0 \quad (12.9)$$

which, in turn, results in:

$$(H_P - H_A) + M(Q_P - Q_A) + RQ_A|Q_A|(x_P - x_A) = 0 \quad (12.10)$$

Similarly, for c-,

$$(H_P - H_B) - M(Q_P - Q_B) + RQ_B|Q_B|(x_P - x_B) = 0 \quad (12.11)$$

The values at points A and B are known. We have two unknowns namely H and Q values at point P and two equations comes from the c+ and c- characteristic lines.

12.3. Charasteristics Method to Solve Saint-Venant Equations

Saint-Venant equations are the governing equations for unsteady open channel flows. The differential form of the Saint-Venant equations can be written in terms of the water depth h for a rectangular channel:

$$\frac{\partial h}{\partial t} + h\frac{\partial V}{\partial x} + V\frac{\partial h}{\partial x} = 0 \quad (12.12)$$

$$\frac{\partial V}{\partial t} + V\frac{\partial V}{\partial x} + g\frac{\partial h}{\partial x} + g(S_f - S_0) = 0 \quad (12.13)$$

For a channel of irregular cross-section these equations can be written as,

$$\frac{\partial h}{\partial t}+\frac{A}{B}\frac{\partial V}{\partial x}+V\frac{\partial h}{\partial x}+\frac{V}{B}\frac{\partial A}{\partial x}h=0 \quad (12.14)$$

$$\frac{\partial V}{\partial t}+V\frac{\partial V}{\partial x}+g\frac{\partial h}{\partial x}+g(S_f-S_0)=0 \quad (12.15)$$

The equations (12.14) and (12.15) are continuity and momentum equations, respectively. In an open channel, a small disturbance propagates towards upstream or downstream. In a rectangular channel, the equation for the celerity of a small disturbance is as follows: $c = (gh)^{0,5}$. For an irregular channel, the speed of a small wave is as follows: $c = (g(A/B))^{0,5}$ where, A is the cross sectional area of the flow in equation and B is the water surface width of the flow, as seen from Figure 12.3. Wave propagation speed c refers to the propagation of a small disturbance, i.e., a small wave, relative to the motion of the fluid. According to an observer standing on the shore, the absolute velocity of the small wave is (V + c) and (V – c), where V is the velocity of the flow (Chanson, 2004).

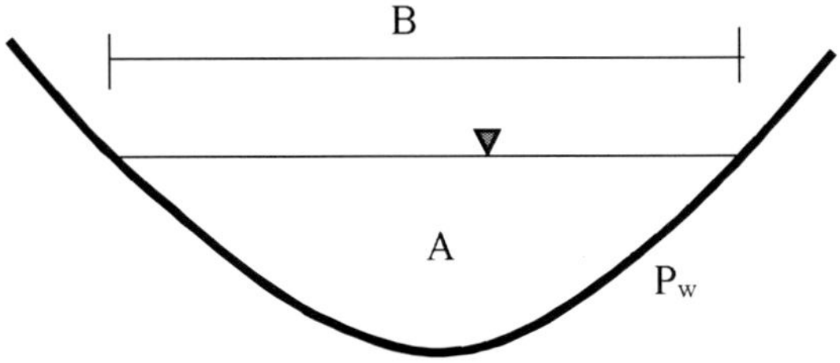

Figure 12.3. Irregular cross section of a channel.

The differentiation of the celerity c, ($c = (gh)^{0,5}$), with respect of time and of space satisfies, respectively:

$$\frac{\partial (c^2)}{\partial t}=2c\frac{\partial c}{\partial t}\approx g\frac{\partial h}{\partial t} \quad (12.16)$$

$$\frac{\partial (c^2)}{\partial x}=2c\frac{\partial c}{\partial x}\approx g\frac{\partial h}{\partial x} \quad (12.17)$$

Replacing the terms $\partial h/\partial t$ and $\partial h/\partial x$ by Equations (12.16) and (12.17), respectively in Equations (12.12) and (12.13) yields:

$$2\frac{\partial c}{\partial t}+2V\frac{\partial c}{\partial x}+c\frac{\partial V}{\partial x}=0 \quad (12.18)$$

$$\frac{\partial V}{\partial t} + 2c\frac{\partial c}{\partial x} + V\frac{\partial V}{\partial x} + g(S_f - S_o) = 0 \qquad (12.19)$$

The addition of Equations (12.18) and (12.19) gives:

$$(\partial/\partial t + (V + c)\,\partial/\partial x)(V + 2c) + g(S_f - S_o) = 0 \qquad (12.20)$$

The subtraction of equation (12.19) from equation (12.18) yields:

$$(\partial/\partial t + (V - c)\partial/\partial x)(V - 2c) + g(S_f - S_o) = 0 \qquad (12.21)$$

Equations (12.20) and (12.21) may be rewritten for forward characteristics as:

$$\frac{d(V + 2c)}{dt} = -g(S_f - S_o) \qquad (12.22)$$

Similarly, for backward characteristics as:

$$\frac{d(V - 2c)}{dt} = -g(S_f - S_o) \qquad (12.23)$$

Along their respective characteristic trajectories for forward characteristic c^+, we can write:

$$\frac{dx}{dt} = V + c \qquad (12.24)$$

Similarly, for backward characteristic c^-, we can obtain:

$$\frac{dx}{dt} = V - c \qquad (12.25)$$

In the characteristics method, the trajectories described above are called characteristic directions or characteristics of the system. As seen from Figure (12.4), for an observer moving forward along the characteristic, Equation (12.24) and Equation (12.22) are valid at any point. Likewise, for an observer moving on the backwards characteristic, Equation (12.25) and Equation (12.24) are satisfied everywhere.

The forward characteristic trajectory is often referred to as c^+ characteristic and positive characteristic. The backward characteristic trajectory is often referred to as c^- characteristic and negative characteristic. For a horizontal ($S_o = 0$), and frictionless channel ($S_f = 0$), the characteristic system of equations, along c^+ characteristic, and along c^- characteristic becomes, respectively: $V + 2c$ = constant, and $V - 2c$ = constant. The constants $(V + 2c)$ and $(V - 2c)$ are referred to the *Riemann invariants*.

If the term $g(S_f - S_o)$ is a constant along the channel and independent of the time, equation (12.21) along c^+ characteristic, and along c^- characteristic becomes, respectively: $d/dt(V + 2c + g(S_f - So)t)$, and $d/dt(V - 2c + g(S_f - S_o)t)$.

That is, the term $(V + 2c + g(S_f - S_o)\ t)$ is a constant along the forward characteristic c^+ and the term $(V - 2c - g(S_f - S_o)\ t)$ is constant along the backward characteristic trajectory c^-.

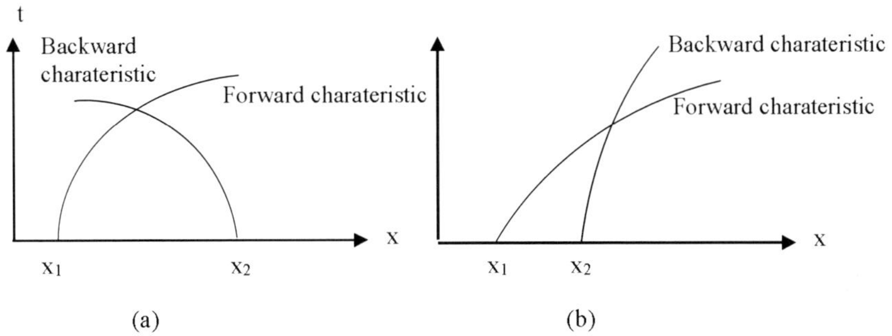

Figure 12.4. Forward and backward characteristics definition sketches: (a) subcritical flow conditions and (b) supercritical flow conditions.

Equations (12.22), (12.23), and (12.24), (12.25) create a system that is identical to Equations (12.12) and (12.23). These four equations describe the typical system of equations that replace Saint-Venant differential equations. The forward (c^+) and backward (c^-) families of the characteristics are shown in Figure 12.4 for the supercritical and subcritical flows in the plane (x, t).

12.4. Solution of Wave Equation

The first order hyperbolic equation (as a linear kinematic model for $\beta = 1$) known as the wave equation or color equation will be solved in detail. The color equation can be written as

$$\frac{\partial f}{\partial t} + c\frac{\partial f}{\partial x} = 0 \qquad (12.26)$$

where f is any function as f = f(x, t) and c is the wave speed.

From the definition of total differential, the following equation can be written as,

$$\frac{df}{dt} = \frac{\partial f}{\partial t} + \frac{dx}{dt}\frac{\partial f}{\partial x} = 0 \qquad (12.27)$$

The first term of the right side, ∂f/∂t, reflects the change in f at the fixed position, whereas (dx/dt)(∂f/∂x) indicates the change caused by the observer moving into a region potentially different f region. It is clear from equation (12.26) that if the observer moves with velocity c.

If c replaced by

$$\frac{dx}{dt} = c \qquad (12.28)$$

$$\frac{df}{dt} = 0 \qquad (12.29)$$

Therefore, f is constant. An observer moving with this special velocity c could measure no changes in f. In this way, Equation (12.29) has been replaced by two ordinary differential equations such as, Equation (12.28) and Equation (12.39).

Equation (12.28) can be written as

$$\int_{t_0}^{t} c dt = \int_{x_0}^{x} dx \qquad (12.30)$$

Integration of Equation (12.30) yields,

$$x = x_0 + c(t - t_0) \qquad (12.32)$$

This equation shows the straight-line families with slope c. These lines called characteristic curves. For $x_0 = 0$ and $t_0 = 0$, the line is called limiting characteristic (See Figure 12.5). The equation of the limiting characteristic is given

$$x = ct \qquad (12.33)$$

If $x_0 = x_0$, $t_0 = 0$ the equation yields

$$x = x_0 + ct \tag{12.34}$$

For $x_0 = 0$, $t_0 = t_0$ the following equation is obtained

$$x = c(t - t_0) \tag{12.35}$$

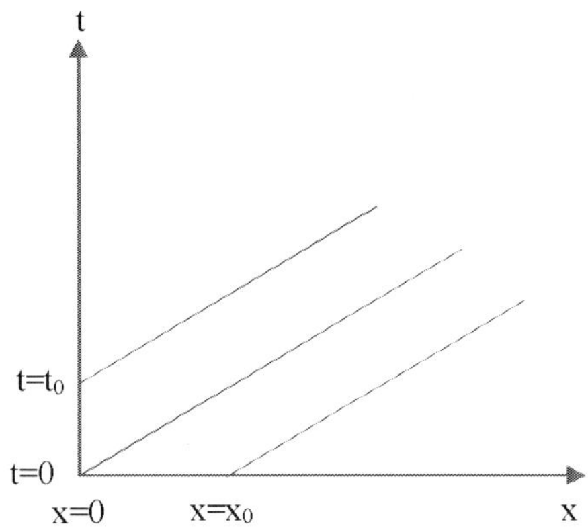

Figure 12.5. Characteristic curves.

The advection equation forms a part of the wave equation as it describes the propagation of a signal along one direction in space.

The equation for the family of parallel characteristics of Equation (12.26), as seen in Figure 12.5. These curves (lines) are called characteristics. Note that at t = 0, $x = x_0$ as initial condition. f(x, t) is constant along this line. f propagates a wave with wave velocity c.

12.5. Kinematic Wave Modeling of Overland Flow

12.5.1. General Remarks

The governing equations of the kinematic wave model for overland flow are the special case of Saint-Venant equations. Any mathematical formulation suitable for the flow on the land surface should be reliable in maintaining the fundamental mass and momentum equations (Ağıralioğlu and Singh, 1980).

The continuity equation in shallow water flows, representing the conservation of mass, can be written as one dimensional,

$$\frac{\partial h}{\partial t} + \frac{\partial q}{\partial x} = i(x, t) \tag{12.36}$$

Or we can write a similar continuity equation as,

$$\frac{\partial A}{\partial t} + \frac{\partial Q}{\partial x} = i(x, t)B \tag{12.37}$$

where q = flow rate per unit width, h = flow depth, x = distance in the flow direction, t = time, *i* = excess rainfall rate, A = cross sectional area of the flow, Q = flow rate, B = wide of the flow.

The momentum equation ($S_0 = S_f$) can be used to generate a flow-resistance equation by ignoring the inertia force and the pressure force as,

$$q = \alpha h^{\beta} \tag{12.38}$$

Using Manning's formula, we can write

$$u = \frac{1}{n} R^{\frac{2}{3}} S_0^{\frac{1}{2}} \tag{12.39}$$

$$q = uh = \frac{1}{n} h^{\frac{2}{3}} h S_0^{\frac{1}{2}} = \frac{1}{n} h^{\frac{5}{3}} s_0^{\frac{1}{2}} \tag{12.40}$$

$$R = h \tag{12.41}$$

$$\alpha = \frac{1}{n} S_0^{\frac{1}{2}} \tag{12.42}$$

$$\beta = \frac{5}{3} \tag{12.43}$$

where *n* is the Manning's roughness coefficient, S_0 is the friction slope.

Substituting Equation (12.38) into Equation (12.36), the kinematic wave equation for the overland flow can be represented as;

$$\frac{\partial h}{\partial t} + \alpha\beta h^{\beta-1} \frac{\partial h}{\partial x} = i(x, t) \tag{12.44}$$

12.5.2. Boundary and Initial Conditions

Consider a watershed flow as seen in Figure 12.6.

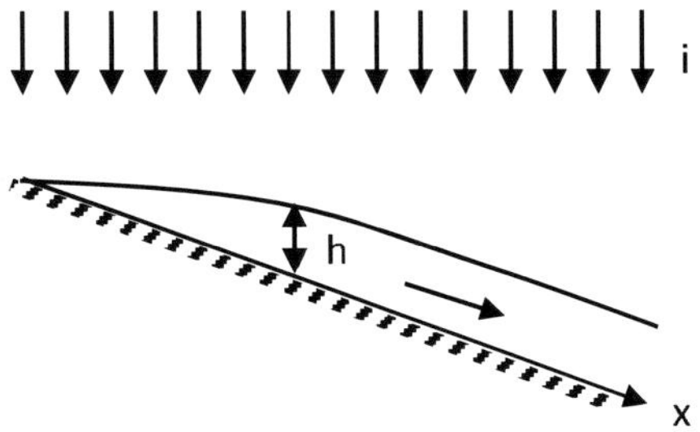

Figure 12.6. Configuration for a watershed flow.

Boundary and initial conditions are given as below.
Boundary condition for x = 0, h = 0.
And initial condition for t = 0, h = 0.
Assumed that rain fall intensity, i = constant.

12.5.3. Analytical Solution Using Characteristics Method

As mentioned before, the method of characteristics is based on converting a partial differential equation to a system of ordinary differential equations (Singh and Ağıralioğlu, 1980; Singh and Ağıralioğlu, 1982).

We can write the total derivatives as;

$$\frac{dh}{dt} = \frac{\partial h}{\partial t}\frac{dt}{dt} + \frac{\partial h}{\partial x}\frac{dx}{dt} \tag{12.45}$$

$$\frac{dh}{dt} = \frac{\partial h}{\partial t} + \frac{\partial h}{\partial x}\frac{dx}{dt} \tag{12.46}$$

Equation (12.46) and Equation (12.48) can be written in matrix form

$$\frac{\partial h}{\partial t} + \alpha\beta h^{\beta-1}\frac{\partial h}{\partial x} = i(x,t) \tag{12.47}$$

$$\frac{\partial h}{\partial t} + \frac{dx}{dt}\frac{\partial h}{\partial x} = \frac{dh}{dt} \tag{12.48}$$

$$\begin{bmatrix} 1 & \alpha\beta h^{\beta-1} \\ 1 & \frac{dx}{dt} \end{bmatrix} \begin{bmatrix} \frac{\partial h}{\partial t} \\ \frac{\partial h}{\partial x} \end{bmatrix} = \begin{bmatrix} i \\ \frac{dh}{dt} \end{bmatrix} \tag{12.49}$$

$$\frac{dt}{dx} = \frac{1}{\alpha\beta h^{\beta-1}} \tag{12.50}$$

$$\frac{dh}{dt} = i \Rightarrow \frac{dh}{dx}\frac{dx}{dt} = i \Rightarrow \frac{dh}{dx} = \frac{i}{\alpha\beta h^{\beta-1}} \tag{12.51}$$

These are the characteristic equations (ODE's). The solution of the characteristic equations is the same as the governing equation, given in Eq. (12.44).

Solution of Eq. (12.50) and Eq. (12.51) only valid on the characteristic curves. To obtain the solution we will distinguish two cases: (1) Equilibrium case, (2) Partial Equilibrium case.

Equilibrium Case: In equilibrium case $t_c < T$ in which T is the rainfall duration, and t_c is the time of concentration.

The characteristic curve through the origin (0, 0) intersects x = L (the downstream boundary before it intersects t = T (the duration of rainfall). This case will give rise to equilibrium hydrograph and is shown in Figure 12.7. The characteristic issuing from the origin is called as the limiting characteristic.

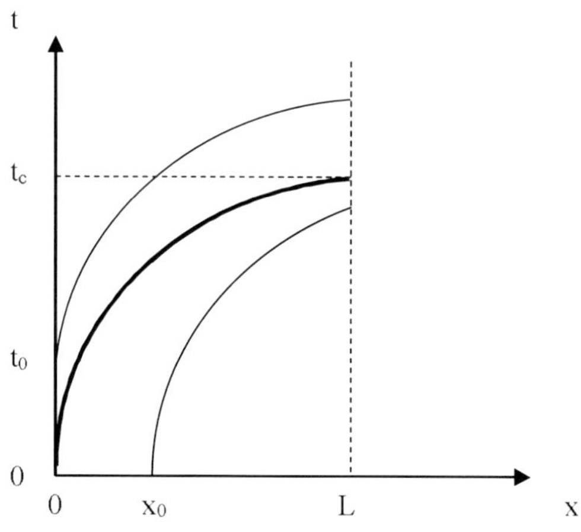

Figure 12.7. Characteristic curves and limiting characteristic.

Partial Equilibrium Case: The characteristic curve through the origin (0, 0) intersects t = T before it intersects the downstream boundary x = L. This case will result in partial equilibrium hydrograph and is shown in Figure 12.8. Thus $t_c > T$ for this case.

The solution to the set two cases will completely characterize the surface runoff hydrograph. We will develop mathematical solutions for the set two cases.

12.5.4. Equilibrium Case

Consider domains as seen in Figure 12.8.

For Domain D_3: From Eq. (12.51) we can write

$$\int \alpha\beta h^{\beta-1} dh = \int i dx \tag{12.52}$$

and

$$\alpha h^{\beta} = ix + c \tag{12.53}$$

For $x = x_0 \rightarrow h = 0$ $(t = 0)$, we obtain

$$0 = ix_0 + c \rightarrow c = -ix_0 \tag{12.54}$$

Therefore

$$\alpha h^{\beta} = ix - ix_0 \tag{12.55}$$

Solving this equation yields

$$h = \left[\frac{i(x - x_0)}{\alpha}\right]^{\frac{1}{\beta}} \tag{12.56}$$

From Equation (12.50) we can write

$$\alpha\beta h^{\beta-1} dt = dx \tag{12.57}$$

$$\int dt = \frac{dx}{[\alpha\beta h^{\beta-1}]} = \int \frac{dx}{\left[\alpha\beta\frac{i(x-x_0)}{\alpha}\right]^{\frac{\beta-1}{\beta}}} \tag{12.58}$$

$$t - t_0 = \frac{1}{\alpha\beta}\left(\frac{i}{\alpha}\right)^{\frac{1-\beta}{\beta}} \int \frac{dx}{(x-x_0)^{\frac{\beta-1}{\beta}}} \tag{12.59}$$

or

$$t - t_0 = \frac{1}{\alpha\beta}\left(\frac{i}{\alpha}\right)^{\frac{1-\beta}{\beta}} \int_{x_0}^{x} (x-x_0)^{\frac{-\beta+1}{\beta}} dx \tag{12.60}$$

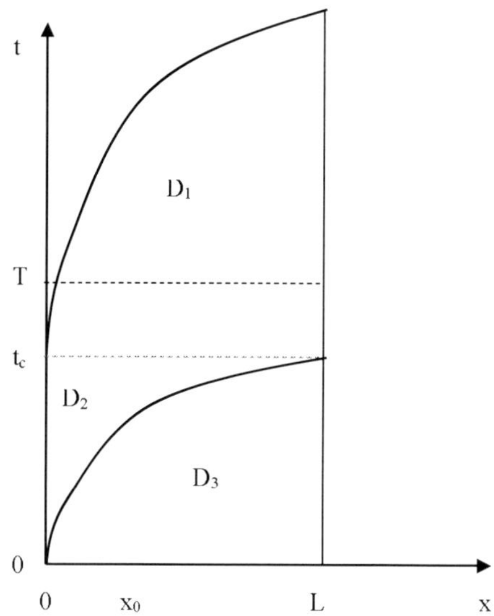

Figure 12.8. Domains for equilibrium case.

For $x = x_0$ and $t_0 = 0$, the solution is given as:

$$t = \frac{1}{\alpha\beta}\left(\frac{i}{\alpha}\right)^{\frac{1-\beta}{\beta}} \int_{x_0}^{x} (x-x_0)^{\frac{1-\beta}{\beta}} dx \tag{12.61}$$

Integrating this equation yields:

$$t = i^{\frac{1-\beta}{\beta}} \left(\frac{1}{\alpha}\right)^{\frac{1}{\beta}} (x-x_0)^{\frac{1}{\beta}} \tag{12.62}$$

For $x_o = 0$ and $x = L$, solution gives the time of concentration ($t = t_c$) as

$$t_c = i^{\frac{1-\beta}{\beta}} \left(\frac{1}{\alpha}\right)^{\frac{1}{\beta}} L^{\frac{1}{\beta}} \tag{12.63}$$

For Domain D_2: For boundary condition $x = 0$, $h = 0$, solving Equation (12.50) yields $c = 0$ and the solution is given by

$$h = \left(\frac{ix}{\alpha}\right)^{\frac{1}{\beta}} \tag{12.64}$$

This equation gives water depth variation in space in Domain D_2. The variation in time can be obtained using Equations (12.58) and (12.50) as

$$dt = \frac{1}{\alpha\beta\left(\frac{ix}{\alpha}\right)^{\frac{\beta-1}{\beta}}} dx \tag{12.65}$$

or $x = 0$, $t = t_0$ the equation gives

$$t = t_o + \frac{1}{\alpha\beta}\left(\frac{i}{\alpha}\right)^{\frac{1-\beta}{\beta}} \int_0^x x^{\frac{1-\beta}{\beta}} dx \tag{12.66}$$

Integrating this yield in

$$t = t_o + i^{\frac{1-\beta}{\beta}} \left(\frac{1}{\alpha}\right)^{\frac{1}{\beta}} x^{\frac{1}{\beta}} \tag{12.67}$$

For $t_o = 0$, Equation (12.67) yields the time of concentration as the same as Equation (12.63) as $t = t_c$.

For Domain D_1: In this domain for $t > T$, $dh/dx = 0$. However, for boundary condition $x = x_0^*$; $h = h_0$.

Therefore, from Equation (12.53) someone obtains

$$c = \alpha h_o^{\beta} - i x_o^* \tag{12.68}$$

Integrating this equation yields:

$$h = \left[\frac{i(x - x_o^*)}{\alpha} + h_o^{\beta}\right]^{1/\beta} \tag{12.69}$$

At the end of Domain D_2 the solution of this equation for $x = x_o^*$ yields

$$h = h_o \tag{12.70}$$

At the end of Domain D_2, for $x = x_o^*$, we can write:

$$h_o = \left(\frac{ix_o^*}{\alpha}\right)^{\frac{1}{\beta}} \tag{12.71}$$

$$\frac{d}{dx}\left(\alpha h^{\beta}\right) = 0 \tag{12.72}$$

$$\alpha h^{\beta} = \text{consant} = c \tag{12.73}$$

$$h^{\beta} = h_o^{\beta} \tag{12.74}$$

$$h = \left(\frac{ix_o^*}{\alpha}\right)^{\frac{1}{\beta}} \tag{12.75}$$

As seen from this equation h is not depend on x, namely is constant. Therefore, it also does not depend on time. Time solution is given as

$$\frac{dt}{dx} = \frac{1}{\alpha\beta h^{\beta-1}} \tag{12.76}$$

$$\int_{t_o}^{t} dt = \int_{x_o^*}^{x} \frac{1}{\alpha\beta}\left(\frac{ix_o^*}{\alpha}\right)^{\frac{1-\beta}{\beta}} \tag{12.77}$$

$$t = T + \frac{1}{\alpha\beta}\left(\frac{ix_o^*}{\alpha}\right)^{\frac{1-\beta}{\beta}}(x - x_o^*) \tag{12.78}$$

12.5.5. Partial Equilibrium Case

In partial equilibrium case, time of concentration, t_c, is greater than rainfall duration ($t_c > T$) (See Figure 12.9).

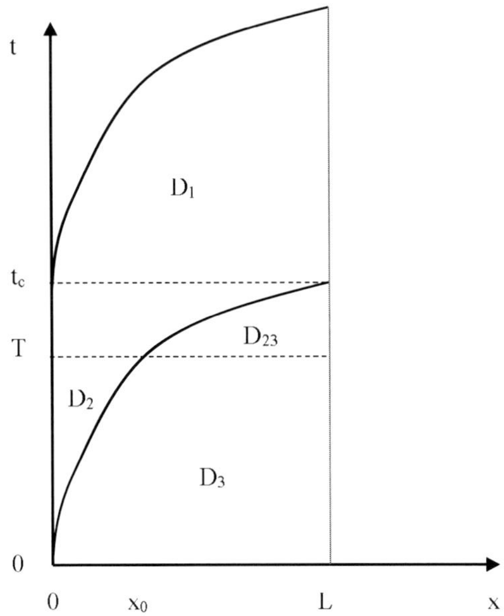

Figure 12.9. Solution Domains for Partial Equilibrium Case.

There may be four domains in this case, as D_1, D_2, D_3 and D_{23}. Solution for domains D_1, D_2 and D_3 are the same for the solution of equilibrium case. Only the solution domain D_{23} will be given here.

Domain D_{23}: The depth and time values at the intersection between domains D_3 and D_{23} can be found by using the solutions of equations 12.56 and 12.62 in Domain D_3.

The solution is formed through the segment for $x = x_o^*$, $h = h_o$ and $t = T$,

$$h_o = \left[\frac{i(x_o^* - x_o)}{\alpha}\right]^{\frac{1}{\beta}} \tag{12.79}$$

$$t_o = T = i^{\frac{1-\beta}{\beta}} \left(\frac{1}{\alpha}\right)^{\frac{1}{\beta}} (x_o^* - x_o)^{\frac{1}{\beta}} \tag{12.80}$$

These two equations show the boundary and initial conditions for Domain D_{23}. In this domain $i = 0$. Therefore, we obtain

$$\frac{dh}{dx} = 0 \tag{12.81}$$

The solutions of this domain are given by

$$h = h_o = \left[\frac{i(x_o^* - x_o)}{\alpha}\right]^{\frac{1}{\beta}} \tag{12.82}$$

$$\int \frac{dt}{dx} = \int_{x_o^*}^{x} \frac{1}{\alpha\beta h^{\beta-1}} \tag{12.83}$$

$$t = T + \frac{1}{\alpha\beta}\left(\frac{i(x_o^* - x)}{\alpha}\right)^{\frac{1-\beta}{\beta}} (x - x_o^*) \tag{12.84}$$

12.6. Applications of Characteristics Method

12.6.1. Solution of Transient Flow Equation

Using the following data, compute the velocity and head at point P. Following data are given at two points A and B along a pipe of diameter 0.25 m carrying water: V_A =4.0 m/s, V_B = 4.2 m/s, H_A = 12 m, H_B = 15 m, Δx = 100 m, Δt = 10 Sec, f = 0.019, c = 1215 m/s. Using the finite difference form of characteristics equations, compute discharge and pressure head at point P (Figure 12.2).

Solution: We can write the characteristic equations in finite difference form for c+, as in Equation (12.10),

$$(H_P - H_A) + M(Q_P - Q_A) + RQ_A|Q_A|(x_P - x_A) = 0$$

Similarly, for C-, as in Equation (12.11)

$$(H_P - H_B) - M(Q_P - Q_B) + RQ_B|Q_B|(x_P - x_B) = 0$$

Substituting given values in these equations for c^+, M = 2523.1 and N = 1.608. The flowrates can be calculated as Q_A = 0.196 and Q_B = 0.206.

$$(H_P - 12) + 2523.1*(Q_P - 0.196) + 1.608*0.196*|0.196|*100 = 0$$

Similarly, for C-,

$$(H_P - 15) + 2523.1*(Q_P - 0.206) + 1.608*0.206*|0.206|*(-100) = 0$$

and,

$H_P + 2523.1\ Q_P = 501.2$

$H_P - 2523.1\ Q_P = -498.35$

Solving these equation yields,

H_p=1.43 m

$Q_P = 0.198\ m^3/s.$

$V_p = 4.03$ m/s

12.6.2. Solution of Saint-Venant Equation

Let's consider a very long channel. The flow values measured at two measurement points in this channel, at t = 0, are given in Table 12.1.

Table 12.1. Given data

Flow properties	Point 1	Point 2
x (km)	1.1	3.5
h (m)	0.77	0.89
V (m/s)	+ 0.70	+ 0.82

Plot the characteristics issuing from each measurement station in the plan (x, t), assuming the characteristic curves are straight. Calculate the location, time and flow properties of the characteristics from the measurement points at the intersection by accepting $S_f = S_0 = 0$.

Solution: The Saint-Venant equations for forward characteristic along and for backward characteristic along become, as Equation (12.22), and Equation (12.23), respectively:

$$\frac{d(V + 2c)}{dt} = -g(S_f - S_o) = 0$$

$$\frac{d(V - 2c)}{dt} = -g(S_f - S_o) = 0$$

For forward characteristic c^+: dx/dt = V + c,
And backward characteristic c^-: dx/dt = V – c.
At t = 0, the basic flow properties are calculated as:

$$c^+ = (gh)^{0.5} = (9.81\times0.77)^{0.5} = 2.75 \text{ m/s}$$

$$c^- = (gh)^{0.5} = (9.81\times0.89)^{0.5} = 2.95 \text{ m/s}$$

$$V_1 + C_1 = 0.70 + 2.75 = 3.45 \text{ m/s}$$

$$V_2 + C_2 = 0.82 + 2.95 = 3.77 \text{ m/s}$$

$$V_1 - C_1 = 0.70 - 2.75 = - 2.05 \text{ m/s}$$

$$V_2 - C_2 = 0.82 - 2.95 = -2.13 \text{ m/s}$$

At t = 0, the calculated basic flow properties are shown in Table 12.2.

Table 12.2. Calculated basic flow properties

Flow properties	Point 1	Point 2
x (km)	1.1	3.5
c(m/s)	2.75	2.95
V + c (m/s)	+ 3.45	+ 3.77
V – c (m/s)	– 2.05	– 2.13
V + 2c(m/s)	+ 6.2	+ 6.72
V – 2c (m/s)	– 4.80	– 5.08

Assuming straight characteristic lines, the equations of the forward characteristic issuing from station 1 and of the backward characteristic issuing from station 2 are respectively:

$$\frac{dx}{dt} = V + c$$

$$\int_{x0}^{x1} dx = \int_{t0}^{t} (V + c)\, dt$$

$$x - x_1 = (V + c)\, t$$

$$x = x_1 + (V_1 + c^+)t$$

Forward characteristic issuing from station 1,

$x = x_1 + 3.45t$

Similarly,

$\int_{x0}^{x1} dx = \int_{t0}^{t} (V - c)\, dt$

Backward characteristic issuing from station 2,

$x = x_2 - 2.13\, t$

$x = 1100 + 3.45\, t$

$x = 3500 - 2.13\, t$

The characteristics intersect at x = 2.58 km for t = 430 s.
The flow properties at intersection satisfy:
Along forward characteristic issuing from station 1,

$V + 2c = V_1 + 2c^+$

$V + 2c = 0.70 + 2\times 3.45 = 7.6$

Along backward characteristic issuing from station 2,

$V - 2c = 0.82 - 2\times 3.77 = -6.72$

It yields: V = +0.44 m/s and c = 3.58 m/s (h = 1.31 m) at the intersection of the characteristics issuing from the two measurement stations assuming $S_f = S_o = 0$.

12.6.3. Calculation of Discharge Hydrograph

For a small watershed calculate the discharge hydrograph for the following data (Ağıralioğlu, 1985a; Ağıralioğlu, 1985b).

Given Data: Watershed length, L = 144 m, watershed slope, $S_o = 0.002$, average width, b = 0.9 m, excess rainfall, i = 19 cm/hour and rainfall duration, T = 8 minutes. Kinematic parameters are taken as, $\alpha = 15.72$ and $\beta = 1.71$.

Solution:

Step 1: Distinguish the equilibrium condition: The equation of concentration time is, as in Equation (12.63),

$$t_c = i^{\frac{1-\beta}{\beta}} \left(\frac{1}{\alpha}\right)^{\frac{1}{\beta}} L^{\frac{1}{\beta}}$$

$$t_c = \left(\frac{0.19}{3600}\right)^{\frac{1-1.71}{1.71}} \left(\frac{144}{15.72}\right)^{\frac{1}{1.71}}$$

$t_c = 218$ seconds = 3.63 minutes < 8 minutes = T.

So, this is an equilibrium case.

Step 2: Solutions for Rising Hydrograph: Using Equation (12.56) and Equation (12.62) we can obtain,

$$h = \left[\frac{i(x - x_o)}{\alpha}\right]^{\frac{1}{\beta}}$$

$$t = i^{\frac{1-\beta}{\beta}} \left(\frac{x - x_o}{\alpha}\right)^{\frac{1}{\beta}}$$

We select x_o between $0 < x_o < 144$, and x = L.

$$Q = B\alpha h^{\beta}$$

$$h = \left[\frac{0.19}{3600}\frac{(144 - x_o)}{15.72}\right]^{\frac{1}{1.71}}$$

$$h = [0.000003357(144 - x_o)]^{0.5898}$$

$$t = \left(\frac{0.19}{3600}\right)^{-0.4152} \left(\frac{144-x_o}{15.72}\right)^{0.5848} = 5.693 \left(\frac{144-x_o}{15.72}\right)^{0.5848}$$

Table 12.3. Calculated values of the rising hydrograph for different x_0

x_o (m)	t (min)	h (cm)	Q (m^3/hour)
0	3.63	1.15	24.62
14.4	3.42	1.08	22.16
28.8	3.19	1.01	19.70
43.2	2.95	0.93	17.24
57.6	2.70	0.85	14.77
72.0	2.42	0.77	12.31
86.4	2.13	0.67	9.85
100.8	1.80	0.57	7.39
115.2	1.42	0.45	4.92
129.6	0.95	0.30	2.46
144.0	0.00	0.00	0.00

We can calculate t and h values for selected values of x_o. Then, it is possible to find Q using the equation $q = \alpha h^{\beta}$. Please note that $Q = qB$ and h is in meters. After the necessary unit conversions Q is obtained in terms of m^3/hour. The results are shown in Table 12.3. Table 12.3 Calculated values of the rising hydrograph for different x_0.

Step 3: Constant Hydrograph: For Domain D_2 as seen in Figure 12.8, from the following equation we obtain h value as for x =144 m, and h =1,15 cm,

$$h = \left(\frac{ix}{\alpha}\right)^{\frac{1}{\beta}}$$

For time values the following equation is valid:

$$t = t_o + i^{\frac{1-\beta}{\beta}} \left(\frac{x}{\alpha}\right)^{\frac{1}{\beta}}$$

Using the given data, we obtain,

$t = t_o$+ 218 seconds (up to 8 minutes)

$t = t_o$+ 3.63 minutes

For selected different t_o, t will be calculated from this equation.

Step 4: Falling Hydrograph: As discussed previously in this chapter, the solutions for water depth and time of concentration at a specific point at Domain D_1 are given by:

$$h = \left(\frac{ix_0^*}{\alpha}\right)^{\frac{1}{\beta}}$$

$$t = T + \frac{1}{\alpha\beta}\left(\frac{ix_0^*}{\alpha}\right)^{\frac{1-\beta}{\beta}}(x - x_0^*)$$

for $0 < x_0^* < 144$ m. For different values of the x_0^* solutions are obtained as

$$h = \left(\frac{0.19}{3600}\frac{x_0^*}{15.72}\right)^{0.5848} = (x_0^* \times 0.000003357)^{0.5848}$$

$$t = 480 + \left(\frac{1}{\alpha}\right)^{\frac{1}{\beta}}\frac{1}{\beta}(i)^{\frac{1-\beta}{\beta}}x_0^{*\frac{1-\beta}{\beta}}(144 - x_0^*) = 480 + 6.972x_0^{*\frac{1-\beta}{\beta}}(144 -$$
$$x_0^*)$$

We can calculate h, t, and Q values for selected x_0^* , and the results are shown in Table 12.4.

Table 12.4. Calculated values of the falling hydrograph for different x_{0*}

x_0^* (m)	t (minute)	h (cm)	Q (m^3/hour)
14.4	12.98	0.30	2.46
28.8	11.32	0.45	4.92
43.2	10.45	0.57	7.39
56.7	9.87	0.67	9.85
72.0	9.42	0.77	12.31
86.4	9.05	0.85	14.77
100.8	8.74	0.93	17.24
115.2	8.47	0.01	19.70
129.6	8.22	1.08	22.16
144.0	8.00	1.15	24.62

The calculated hydrograph is shown in Figure 12.10.

The overall hydrograph resulted from the excess rainfall i = 19 cm/hour that lasts T = 8 minutes is shown in Figure 12.11.

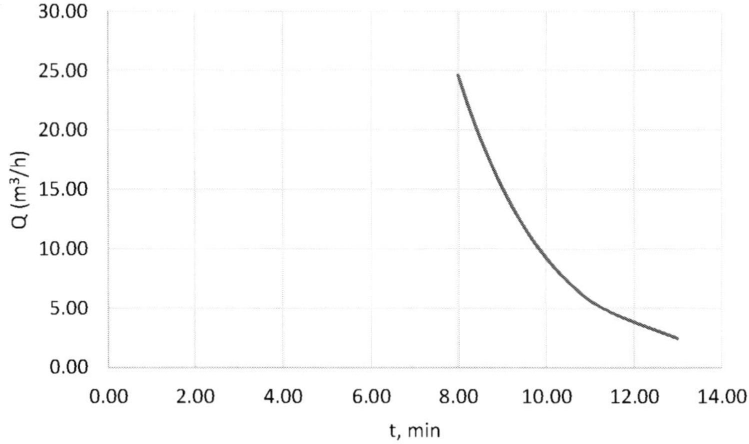

Figure 12.10. Calculated falling hydrograph.

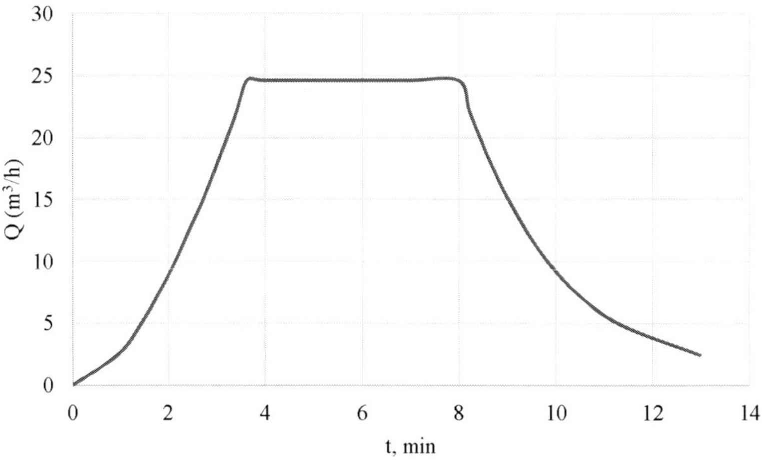

Figure 12.11. Calculated overall hydrograph.

12.6.4. Solution of Advection Equation

Another partial differential equation solution utilizing the method of characteristic is shown below. Let's consider the following equation,

$$\alpha\frac{\partial u}{\partial x}+\frac{\partial u}{\partial t}=0$$

where α is constant and u is a function of x and t.

Solution: This linear first-order partial differential equation must be transformed into an ordinary differential equation along the relevant characteristics curve, i.e., something of the form

$$\frac{d}{ds}u\big(x(s), t(s)\big) = F(u, x(s), t(s))$$

where (x(s), t(s)) is a characteristic line. First, let's find

$$\frac{d}{ds}(u(s), t(s)) = \frac{\partial u}{\partial x}\frac{dx}{ds} + \frac{\partial u}{\partial t}\frac{dt}{ds}$$

by the exact differential. Now, if we replace ($dx/ds = \alpha$) and ($dt/ds = 1$) we obtain

$$\frac{d}{ds}\big(u(s), t(s)\big) = \alpha\frac{\partial u}{\partial x} + \frac{\partial u}{\partial t}$$

This is the left-hand side of the partial differential equation from which we started. Thus

$$\frac{dy}{ds}u = \alpha\frac{\partial u}{\partial x} + \frac{\partial u}{\partial t} = 0$$

As a result, along the characteristic line (x(s), t(s), the original partial differential equation $u_s = F(u, x(s), t(s)) = 0$ is transformed into the ordinary differential equation us = F (u, x(s), t(s)) = 0. That is, the solution is constant along the characteristics. As a result, $u(x_s, t_s) = u(x_0, 0)$, where (x_s, t_s) and $(x_0, 0)$ have the same characteristic. To obtain the general solution, it is sufficient to solve the characteristic system of ordinary differential equations:

$$\frac{dt}{ds} = 1$$

Letting $(t_0) = 0$, we know $t = s$,

$$\frac{dx}{ds} = \alpha$$

Letting $x(0) = x_0$, we know

$$x = \alpha s + x_0 = \alpha t + x_0$$

$$\frac{du}{ds} = 0$$

u is a constant.
Thus,

$$x = x_0 + \alpha t$$

u is a constant.

The characteristic lines in this case are straight lines with slope α, and the value of *u* remains constant along any characteristic line, as shown in Figure 12.5.

12.6.5. Calculation of the Time of Concentration for a Diverging Surface

Find the time of concentration for a diverging surface as a parking spot for the following data. Compare the results with the rectangular geometry for the same condition (Ağıralioğlu, 1987).

Given data: Rainfall intensity = 5 cm/hour, Degree of divergence, c = 0.20, Slope = 0.005, Length = 100 m, Manning roughness coefficient = 0.028.

Solution: For a diverging surface, the time of concentration is given by

$$t_c = \left(\frac{1-c}{i}\right)^{0.4} \left(\frac{nL}{\sqrt{S_0}}\right)^{0.6}$$

$$t_c = \left(\frac{1-0.2}{\frac{0.05}{3600}}\right)^{0.4} \left(\frac{0.028 \times 100}{\sqrt{0.005}}\right)^{0.6} = 729.18 \text{ s} = 12.5 \text{ Min.}$$

For rectangular geometry: $t_c = i^{\frac{1-\beta}{\beta}} \left(\frac{1}{\alpha}\right)^{\frac{1}{\beta}} L^{\frac{1}{\beta}}$

$$t_c = L^{0,6} n^{0,9} / (S_o{}^{0,3} i^{0,4})$$

$$t_c = 797.56 \text{ s} = 13.29 \text{ min.}$$

Diverging geometric model gives smaller time of concentration than the rectangular one.

12.6.6. Calculation of Rising Hydrograph

For equilibrium case, the equation of rising hydrographs is give as

$$h = \left[\frac{i(x-x_0)}{\alpha}\right]^{\frac{1}{\beta}}$$

$$t = i^{\frac{1-\beta}{\beta}}\left(\frac{x-x_0}{\alpha}\right)^{\frac{1}{\beta}}$$

Find the hydrograph values at the length of 200 m for a rectangular watershed for the following data.

Given data: Kinematic parameters as $\alpha = 1$, $\beta = 5/3$, Rainfall intensity: i = 0.20 m/hour, and Length and width of the watershed are L = 200 m and B = 0.9 m, respectively.

Solution: We select $x_o = 0$, $x_o = 50$ m, $x_o = 100$, $x_o = 150$ m, and $x_o = 200$ m.

Calculation is made using the following table ($\beta = 1.67$).

Table 12.5. Calculated values of the rising hydrograph for different x_0

x_o (m)	t (sec)	t (min.)	h (m)	$q = \alpha h^{\beta}$	
				m^2/s	m^2/h
(1)	(2)	(3)	(4)	(5)	(6)
0	1209.83	20.16	0.06721	0.0110	40
50	1018.03	16.96	0.0565	0.0082	30
100	798.19	13.30	0.044	0.00549	20
150	526.61	8.77	0.029	0.00274	10
200	0	0	0	0	0

Using values of 3. and 6. columns the rising hydrograph is plotted in Figure 12.12.

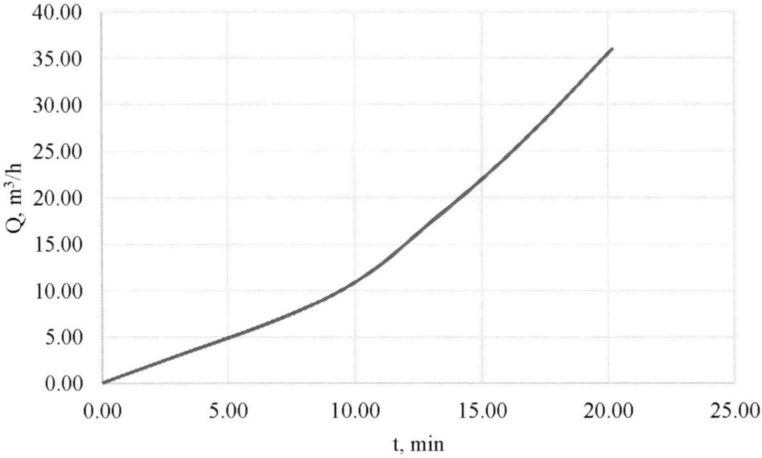

Figure 12.12. Calculated rising hydrograph.

12.6.7. Calculation of Time of Concentration

Find the time of concentration for the following data for a watershed.

Rainfall intensity: $i = 0.12$ cm/hr, Bottom slope: $S_o = 0.04$, Watershed length: $L = 100$ m, and Manning's friction coefficient: $n = 0.035$.

Solution: Time of concentration can be obtained,

$$t_c = L^{0.6} n^{0.9} / (S_o{}^{0.3} i^{0.4})$$

$$i = \frac{0.12}{100}\frac{1}{3600} = 3.33x10^{-7} m/s$$

$$t_c = \frac{(100)^{0.6}(0.035)^{0.9}}{(3.33x10^{-7})^{0.4}(0.04)^{.3}} = 797s = 13.3 \text{ minutes.}$$

12.6.8. Calculation of Concentration Time Using Kinematic Wave Model

For an overland flow, find the time of concentration for the following data: $n = 0.05$, $S_0 = 0.04$, $i = 20$ mm/hr, $L = 500$ m.

Solution: The following formulas are derived in subchapter 12.5:

$$t_c = \left[L^{1/\beta}\, i^{\frac{1-\beta}{\beta}}\right] / \alpha^{1/\beta}$$

$$q = \alpha h^{\beta}$$

$$q = \frac{Q}{B} \text{ and } A = B.h$$

$$Q = \frac{1}{n} R^{2/3} {S_0}^{1/2} A$$

$$R \cong h$$

$$q = \frac{1}{n} {S_0}^{1/2} h^{5/3}$$

$$\alpha = \frac{1}{n} {S_0}^{1/2} = \frac{1}{0.05} (0.04)^{1/2} = 4$$

$$\beta = 5/3$$

And time of concentration,

$$t_c = \left[\frac{500\left(\frac{20}{3600\times1000}\right)^{1-5/3}}{4}\right]^{3/5} = 2292 \text{ seconds} = 38.22 \text{ minutes}$$

References

Ağıralioğlu N. (1985a) 'A Comparison of Water Lag Times for Converging and Plane Surfaces,' *Nordic Hydrology*, 16 (3), pp. 169-176.

Ağıralioğlu, N. (1985b) 'Analytical Solution of Leaf-Shaped Basin Flow,' *Hydrological Sciences Journal-Journal Des Sciences Hydrologiques*, 30 (3), pp. 407-424.

Ağıralioğlu, N. (1987) 'A Dimensionless Solution of Lag Time for Diverging Surface,' *Nordic Hydrology*, 18 (1), pp. 59-62.

Ağıralioğlu, N., and Singh, V. P. (1980). *A Mathematical Investigation of Diverging Flow: 2. Numerical Solutions and Applications*, Technical Report, MSSU-EIRS-80-4, Engineering and Industrial Research Station, Mississippi State University.

Ali, Mostafa. 2014. *Lecture Note on Fundamentals of Computational Hydraulics*. Bangladesh University of Engineering and Technology.

ASCE (Task Committee on Glossary of Hydraulic Modeling Terms of the Committee on Research of the Hydraulics Division). (1982) 'Modeling Hydraulic Phenomena: A Glossary of Terms,' *Journal of the Hydraulics Division*, 108 (7), pp. 845-852.

Asmar, N. H., 2004. *Students Solutions Manual Partial Differential Equations with Fourier Series and Boundary Value Problems*, Second Edition, University of Missouri.

Bermudez, A. and Vazquez, M. E. (1994) 'Upwind methods for hyperbolic conservation laws with source terms,' *Computers and Fluids*, 23(8), 1049–1071.

Causon, D. M., and Mingham, C. G. (2010) *Introductory Finite Difference Methods for Partial Differential Equations*, Ventus Publishing,

Chanson, H. (2004) *Environmental Hydraulics of Open Channel Flows*. 2nd ed. Oxford, UK: Elsevier Butterworth-Heinemann.

Chasnov, J. R., (2019) *Differential Equations for Engineers*, Lecture Notes, The Hong Kong University of Science and Technology.

Chapra, S. C., and Canale, R. P. (1986) *Numerical Methods for Engineers*, New York: McGraw Hill.

Chaudhry, M. H. (2008) *Open-channel Flow*. New York: Springer.

Chow, V. T. (1959) *Open-channel Hydraulics*. New York: McGraw-Hill Inc.

Cunge, J. A., Holly, F. M. Jr. and Verway, A. (1980) *Practical Aspects of Computational River Hydraulics*. London: Pitman Publishing Limited.

Dubin. J. R. (1999) 'On gradually varied flow profiles in rectangular open channels,' *Journal of Hydraulic Research*, 37, pp 99 - 106.

Fenton, J. (2010) *Computational Hydraulics*, TU Wien, Institut für Wasserbau und Ingenieurhydrologie Karlsplatz, Wien.

Franke, O. L., Reilly, T. E., and Bennett, G. D. (1987) *Definition of Boundary and Initial Conditions in The Analysis of Saturated Ground-Water Flow Systems-An Introduction*: U.S. Geological Survey Publications, TWRI, 3-B5, P. 22.

Gary, J. (1969) *The Numerical Solution o Partial Differential Equations*, National Center for Atmospheric Research, Boulder, Colorado.

Gentry, R., Martin, R., and Daly, B. (1966) 'An Eulerian differencing method for unsteady compressible flow problems.' *J. Comput. Phys.* 1, 87-118.

Golub, G. H., and Ortega, J. M. (1993). *Scientific Computing and Differential Equation: An Introduction to Numerical Methods*, Academic Pres.

Hanson, H., and Kraus, N. C. (1989) GENESIS: *Generalized model for simulating shoreline change, Report 1: Technical Reference*. Tech. Rep. CERC-89-19, U.S. Army Engineer Waterways Experiment Station, Coastal Engineering Research Center, Vicksburg, MS.

Hoffman, J. D. (2001) *Numerical Methods for Engineers and Scientists*, Second Edition, Revised and Expanded, Basel: Marcel Dekker Inc.

Jenkins, W. M., Coulthard, Jm., Jesus, E. C. De. (1983) *BASIC Computing for Civil Engineers*, Van Nostrand Reinhold.

Kamphuis, J. W. (2000) *Introduction to Coastal Engineering and Management*, World Scientific Publishing Co. Pte. Ltd.

Koutitas, C. G., (1983) *Elements of Computational Hydraulics*, Pentech Pres.

Kumar, M. S. M. (2010) *Computational Hydraulics*, India.

Liggett, J. A. (1975) *Basic equations of unsteady flow. In: Unsteady Flow in Open Channels (K. Mahmood and V. Yevjevich, Eds.)*. Vol. 1. Water Resources Publications, Fort Collins, Colorado, USA. pp. 485–508.

Liggett, J. A. and Woolhisher, D. A. (1967) 'Difference solutions of shallow water equations.' *J. Engineering Mechanics Division,* 93(EM2), 39–71.

Liggett, J. A. and Cunge, J. A. (1975) *Numerical methods of solution of the unsteady flow equations. In: Unsteady Flow in Open Channels (K. Mahmood and V. Yevjevich, Eds.)*. Vol. 1. Water Resources Publications, Fort Collins, Colorado, USA. pp. 485–508.

Lynch, D. R. (2005) *Numerical partial differential equations for environmental scientists and engineers: A First Practical Course*, Springer.

Massau, J. (1900) *MeMoire Sur L'InteGration Graphique Des eQuations Aux DeRiveEs Partielles*. Clear Prints.

McDonough, J. M. (2008) *Lectures on computational numerical analysis of partial differential equations*. Departments of Mechanical Engineering and Mathematics University of Kentucky, 169 p.

Meleshko, S. V., (2005) *Methods for Constructing Exact Solutions of Partial Differential Equations, Mathematical and Analytical Techniques with Applications to Engineering*. Springer.

Miller, J. G. (1984) *Basic concepts of kinematic-wave models*. Geological Survey Professional Papers 1302, US Government Printing Office, Washington, 29 p.

Mohan, M. S. (2010) *Computational Hydraulics*, Department of Civil Engineering, Indian Institute of Science.

Morton, K. W. and Mayers, D. F. (2005) *Numerical Solution of Partial Differential Equations: An Introduction*. Second Edition, Cambridge University Press.

Notdström, J. (2015) 'Well Posed Problems and Boundary Conditions in Computational Fluid Dynamics,' American Institute of Aeronautics and Astronautics, Conference Paper, P. 1-14.

Popescu, I. (2014) *Computational Hydraulics: Numerical Methods and Modelling*. IWA Publishing.

Ponce, V. M. and Simons, D.B. (1977) 'Shallow wave propagation in open channel flow,' *J. of Hydraulics Division,* 103(12), 1461–1476.

Rezzolla, L. (2011) *Numerical Methods for the Solution of Partial Differential Equations*. Lecture Notes for the COMPSTAR School on Computational Astrophysics, 8-13.

Roache, P. J. (1979) *Computational Fluid Dynamics*, Hermosa Publishers.

Salamanzadeh, M., Torfi, S. (2011) 'Analysis of Surge in Pipelines Systems by Characteristics Method,' *International Journal of Mechanics*, 2(5), pp. 83-90.

Santiago-Collazo. F. L., Araya. S. A. (2014) 'Computation of Gradually Varied Flow in a Channel System,' Conference Paper Conference: Puerto Rico Louis Stoke Association for Minority Participation, At UPR- Rio, Piedras, 57 p.

Shieh. J. C. (2009) *Fundamentals of Fluid Mechanics: Chapter 10. in Open Channels Flow*. Department of Bio-Industrial Mechatronics Engineering. National Taiwan University.

Singh V. P, and Ağıralioğlu, N. (1982) 'Lag Time for Diverging Overland-Flow,' *Nordic Hydrology*, 13 (1), pp. 39-48.

Singh, V. P. and Ağıralioğlu, N. (1980) *A Mathematical Investigation of Diverging Flow: Analytical Solutions*. Technical Report, MSSU-EIRS-80-3, Engineering and Industrial Research Station, Mississippi State University, 175 p.

SPM. (1984). *Shore Protection Manual: Volume I and Volume II*. USACE, Department of the Army, U.S. Corps of Engineers, Washington, DC 20314.

Stelling, G. S., and Booij, N. (1999) *Computational modelling flow and transport*. Lecture Notes, Delft University of Technology, Faculty of Civil Engineering and Geosciences.

Szymkiewicz, R. (2010) *Numerical Modeling in Open Channel Hydraulics, Faculty of Civil and Environmental Engineering*. Gdansk University of Technology, Poland, : Springer Science+Business Media B. V.

Szymkiewicz, R., Huang, S., Szymkiewicz, A. (2016) *Introduction to Computational Engineering Hydraulics*. Gdańsk University of Technology Publishing House.

Tey, W. Y, Asako, Y., Sidik, N. A. C., and Rui-Zher, G. (2017) 'Governing Equations in Computational Fluid Dynamics: Derivations and A Recent Review,' *Progress in Energy and Environment*, 1, 1 – 19.

Timoshenko, S., and Woinowsky-Krieger, S. (1987) *Theory of Plates and Shells, (2nd Edition)*. McGraw-Hill.

Yanovsky, I. (2005) *Partial Differential Equations: Graduate Level Problems and Solutions*. Available at: https://www.math.ucla.edu/~yanovsky/handbooks/PDEs.pdf. (Accessed: 09 August 2023).

About the Authors

Necati Ağıralioğlu

Necati Ağıralioğlu's journey through the realm of civil engineering has been one marked by unwavering dedication and scholarly achievement. Graduating with both a Bachelor's and a PhD in Civil Engineering from Istanbul Technical University, in 1968 and 1977, respectively, he embarked on a path of academic excellence that continues to this day.

Currently serving as a respected Professor at Antalya Bilim University, Dr. Ağıralioğlu has etched his name in the annals of academia with a remarkable record of accomplishments.

Dr. Ağıralioğlu's contributions extend far beyond the classroom. With a literary prowess that has yielded an impressive collection of 10 books in the field of water engineering, he has illuminated the path for countless scholars and practitioners. Moreover, his mentorship has guided and shaped the careers of more than 50 graduate students, a testament to his commitment to fostering the next generation of talent.

In the arena of research, Dr. Ağıralioğlu's involvement in numerous scientific projects centered on water engineering underscores his commitment to addressing critical issues facing our global water resources. His substantial body of work comprises over 100 publications, encompassing both journal articles and conference papers, reflecting his enduring quest for knowledge dissemination.

At present, Dr. Ağıralioğlu's primary focus lies in enhancing the quality of engineering education, a pursuit that carries profound implications for the future of the profession. Simultaneously, he continues to explore water resource-related challenges, delving into the intricacies of issues such as floods and droughts, with an unwavering commitment to finding sustainable solutions.

Necati Ağıralioğlu's life and work encapsulate the essence of a lifelong dedication to civil engineering, education, and research. His profound impact on the field and his unwavering commitment to addressing critical global challenges serve as an enduring legacy.

Mehmet Özger

Dr. Mehmet Özger is a distinguished scholar with a strong academic background in Civil Engineering. He earned his MSc. and PhD. degrees in Civil Engineering from the prestigious Istanbul Technical University, establishing a solid foundation for his career. Currently, he holds the esteemed position of Professor at the very same institution, where he continues to make remarkable contributions to the field.

Dr. Özger's expertise lies in the domain of water engineering, and he has actively engaged in numerous pioneering scientific projects in this field. His prolific research career has yielded a wealth of knowledge, with a substantial portfolio of over 100 scholarly publications encompassing journals and conference papers.

With a passion for environmental sustainability, Dr. Özger has dedicated his recent efforts to the vital issue of sustainable water drainage within urban areas. This research avenue explores innovative solutions to effectively manage water resources in increasingly complex city environments. Furthermore, he has delved into the application of cutting-edge modeling techniques to address critical water-related challenges, including flood mitigation, drought resilience, and optimizing water supply systems.

In sum, Dr. Mehmet Özger is a prominent figure in the realm of Civil Engineering, a dedicated educator, and a pioneering researcher whose work is instrumental in shaping the future of water engineering and urban sustainability.

Index

S

T

U

V

W